Nanochemistry for Chemistry Educators

Advances in Chemistry Education Series

Editor-in-chief:
Keith S. Taber, *University of Cambridge, UK*

Series editors:
Avi Hofstein, *Weizmann Institute of Science, Israel*
Vicente Talanquer, *University of Arizona, USA*
David Treagust, *Curtin University, Australia*

Titles in the Series:
1: Professional Development of Chemistry Teachers: Theory and Practice
2: Argumentation in Chemistry Education: Research, Policy and Practice
3: The Nature of the Chemical Concept: Re-constructing Chemical Knowledge in Teaching and Learning
4: Creative Chemists: Strategies for Teaching and Learning
5: Engaging Learners with Chemistry: Projects to Stimulate Interest and Participation
6: The Johnstone Triangle: The Key to Understanding Chemistry
7: Problems and Problem Solving in Chemistry Education
8: Teaching and Learning in the School Chemistry Laboratory
9: Nanochemistry for Chemistry Educators

How to obtain future titles on publication:
A standing order plan is available for this series. A standing order will bring delivery of each new volume immediately on publication.

For further information please contact:
Book Sales Department, Royal Society of Chemistry, Thomas Graham House, Science Park, Milton Road, Cambridge, CB4 0WF, UK
Telephone: +44 (0)1223 420066, Fax: +44 (0)1223 420247,
Email: booksales@rsc.org
Visit our website at www.rsc.org/books

Nanochemistry for Chemistry Educators

By

Riam Abu Much
The Academic Arab College for Education, Israel
Email: riama@arabcol.ac.il

Kurt Winkelmann
Florida Institute of Technology, USA
Email: kwinkelmann@valdosta.edu

and

Muhamad Hugerat
The Academic Arab College for Education, Israel
Email: muha4@arabcol.ac.il

ROYAL SOCIETY OF CHEMISTRY

Advances in Chemistry Education Series No. 9

Print ISBN: 978-1-78801-886-9
PDF ISBN: 978-1-83916-477-4
EPUB ISBN: 978-1-83916-478-1
Print ISSN: 2056-9335
Electronic ISSN: 2056-9343

A catalogue record for this book is available from the British Library

The Royal Society of Chemistry is a charity, registered in England and Wales, Number 207890, and a company incorporated in England by Royal Charter (Registered No. RC000524), registered office: Burlington House, Piccadilly, London W1J 0BA, UK, Telephone: +44 (0) 20 7437 8656.

For further information see our website at www.rsc.org

Printed in the United Kingdom by CPI Group (UK) Ltd, Croydon, CR0 4YY, UK

Preface

Chemistry is central to understanding all science, including the broad field of nanotechnology. Nanoscale chemistry is at the forefront of modern science in the 21st century. Its applications are transforming society: modern electronics and consumer goods make our daily lives easier and new medical treatments and cleaner energy improve our long-term prosperity. These technological advances raise questions about how industry and government should implement nanochemical-based technologies in a safe and fair manner.

Providing future societal benefits from scientific and technological advances requires our action now. Teachers must learn how to teach nanochemistry to students in middle and high school in order to create an informed public and scientifically knowledgeable workforce. A better understanding of nanoscale phenomena and applications enables citizens to navigate the science-based issues that relate to their everyday lives and to intelligently question the ethical and societal implications of these revolutionary technologies. Applications of nanochemistry in medicine, materials, energy, biotechnology, and other industries will create millions of jobs for our scientifically-minded students.

Although nanochemistry is a relatively new field of science, this book demonstrates that teaching nanochemistry requires the same proven techniques that teachers already use for engaging and inspiring students. We also encourage teachers to learn about emerging pedagogies within science education and apply them to teaching nanochemistry. Students can learn about nanochemistry through hands-on activities, including those that do not require any laboratory facilities. Curriculum materials highlight ways to connect the underlying chemical concepts to applications that interest students. Citations of primary sources allow readers to dig deep into

Advances in Chemistry Education Series No. 9
Nanochemistry for Chemistry Educators
By Riam Abu Much, Kurt Winkelmann and Muhamad Hugerat

Published by the Royal Society of Chemistry, www.rsc.org

the original education research to learn more about the background of each topic. Researchers and policy makers have suggested that there is a need to update science curricula by adding new topics such as nanochemistry. However, teachers have responded that there is no room for new topics in their already rigorous curriculum. To address these concerns, nanoeducation researchers have discovered creative ways to insert nanochemistry in individual lessons throughout the curriculum or as a standalone unit.

The authors of this book are scientists and researchers who have studied nanochemistry and nanochemistry education for many years. We are motivated to write this book because we believe that middle and high school teachers need more support and encouragement for introducing their students to nanochemistry. Teachers would use the significant amount of nanochemistry curricular materials that are available if they better understand the relevance of nanochemistry, knew about helpful educational resources, and had a plan for integrating topics into their existing lessons. We summarize many scientific and non-scientific issues and topics at a suitable level for current and future teachers who have no previous knowledge about nanoscience. In addition, our book offers ideas for how to include nanoscience concepts and applications when teaching chemistry. The information in this book may be of interest to students with an interest in nanochemistry as well.

We begin with background about nanochemistry, then explain why and how teachers can incorporate this topic into their lessons. The authors refer the reader to other chapters for additional information about a topic when appropriate. Chapter 1 presents the science of nanochemistry and its applications. This is suitable for teachers who are new to this subject or need to refresh their knowledge. The chapter describes the basic concepts and scientific theories that explain the behavior of materials, structures, and systems at the nanoscale. Chapter 2 presents a broad literature review about the current awareness and attitudes of teachers towards nanochemistry. Chapter 3 gives an overview of how teachers can share nanochemistry with their students. Chapter 4 describes strategies drawn from the STEM educational research literature to explain nanoscale chemistry to students in different educational settings, both formal and informal. It also explains how teachers can overcome common challenges that they may face. Chapters 5 reviews the many activities that teachers can lead that will engage their students in learning about nanochemistry. In Chapter 6, we explore how teachers can present nanochemistry within a context that relates to students' interests. Chapter 7 provides an example of a nanochemistry teaching unit about nanoliposomes and their use as vehicles to deliver pharmaceutical drugs in the human body. The chapter includes additional scientific background information about the chemical structure of nanoliposomes and its relationship to the chemical structure of the cell membrane. The book concludes with Chapter 8 discussing how teachers can present the ethical issues related to nanochemistry and its effect on society, the environment, and safety.

This book will be useful to pre-service and current teachers with an interest in updating their curriculum with a cutting edge, scientific topic. Throughout the book, we summarize the current state of the education and nanochemistry research literature as well as provide practical advice for teaching this important subject. Above all, we hope that our book inspires teachers to learn more about nanochemistry and helps them inspire their own students to do the same.

Riam Abu Much, Kurt Winkelmann, and Mohamad Hugerat

Author Biographies

Riam Abu Much serves as the Head of the Chemistry Department at The Academic Arab College for Education in Israel, Haifa. She is the former Head of Science Department. She also serves as a Senior Lecturer at the same college. She earned her PhD degree in chemistry (Nanoscience and Nanotechnology specialization) from Bar-Ilan University, she also received her MSc and BSc in chemistry from the Hebrew University of Jerusalem. In 2008–2009 she was a post-doctoral fellow in chemistry at Bar-Ilan University. Riam's research focuses on nanoscience and nanotechnology. During her PhD study, she worked on developing metals and magnetic materials at the nanoscale using Ultrasound Irradiation. In addition, she served as a teaching assistant at Bar-Ilan University. At the end of 2010, her research expanded to include chemical education with an emphasis on nanoeducation. Riam served previously as a scientific member in *SALiS* (Student Active Learning in Science) and *ARTiST* (Action Research to Innovate Science Teaching) projects funded by EU. She has published several articles in different international journals. She has organized several international conferences and presented her research achievements in several conferences around the world.

Dr. Kurt Winkelmann serves as the Head of the Department of Chemistry at Valdosta State University in Valdosta, Georgia, United States. He studied photochemical reaction kinetics and nanotechnology as a PhD student at Auburn University and as a postdoctoral researcher at Northwestern University. He continued this research as a faculty member at Florida Tech for about 20 years. His first experience teaching chemistry piqued his interest in understanding how students learn chemistry. Since then, his research has expanded to include chemical education and nanotechnology education

Advances in Chemistry Education Series No. 9
Nanochemistry for Chemistry Educators
By Riam Abu Much, Kurt Winkelmann and Muhamad Hugerat

Published by the Royal Society of Chemistry, www.rsc.org

research. He has a strong interest in understanding student learning in the chemistry laboratory and classroom. Current and past projects include developing and accessing new academic courses, creating and implementing new lab experiments for first-year students, and using digital learning environments such as virtual worlds for chemistry lab, among others.

Muhamad Hugerat is a Professor of Science Education. He has a PhD in Physical Chemistry from The Hebrew University – Jerusalem. He is former Vice-President of the Academic Arab College for Education (AACE) in Israel – Haifa, former Dean of the Science Education Department, Former Dean of the Continuing Education Department, Head of the *SALiS* (Student Active Learning in Science) Center, and Head of *ARTIST* (Action Research to Innovate Science Teaching) Center. He is editor of "*DARUNA*" – a journal of the AACE. He is a researcher in the Institute of Science Education – The Galilee Society, Shefar-Am. He has conducted research in many areas of science education (*e.g.*, learning environment, laboratory work, science education in all level ages, microscale and nano laboratories, teaching science through research, argumentation and sustainability in science education, history and philosophy of science, and green and environmental science education). He has published several articles in different international journals. He has written many teaching books in science education for the schools of Israel. He has been involved in organising and speaking at many national and international Science Education conferences.

Contents

Advances in Chemistry Education Series No. 9
Nanochemistry for Chemistry Educators
By Riam Abu Much, Kurt Winkelmann and Muhamad Hugerat

Published by the Royal Society of Chemistry, www.rsc.org

CHAPTER 1

Nanotechnology and Chemistry: The Unseen Scale with Magnificent Impact

1.1 Introduction

"There's Plenty of Room at the Bottom."

This was the title of physicist Richard Feynman's talk on December 29, 1959, at the Annual Meeting of the American Physical Society at The California Institute of Technology (Caltech). In that talk, Feynman asked many questions widely believed to be illogical at the time. Feynman believed that if we were able to control matter at the atomic or molecular scale, we could achieve many practical advances that could revolutionize technology and production processes.[1]

Feynman posed many questions related to very small-scale systems and their enormous usefulness. For example, he discussed how the small size of living cells aids in executing important functions for living organisms. Feynman proposed an idea that if we could have the ability to manipulate things as small as atoms and molecules, that these manipulations could be used in a limitless range of technological applications. He imagined that if we could control small-scale substances, we could create materials with new chemical and physical properties for new applications. Essentially, Feynman focused his talk on the idea of nanoscience and nanotechnology, but he never explicitly stated it.[1]

Advances in Chemistry Education Series No. 9
Nanochemistry for Chemistry Educators
By Riam Abu Much, Kurt Winkelmann and Muhamad Hugerat

Published by the Royal Society of Chemistry, www.rsc.org

1.1.1 What Is Nano and Why Is It Important?

Linguistically, the word "nano" is derived from the Greek word pertaining to dwarf, which refers to very small things. One nanometre (nm) is equal to one billionth of a meter (0.000 000 001 m). Here are some relative comparisons which describe the nanometre scale:

- If a one mm section on a ruler is divided into one million equal parts, this is one nanometre in length.
- The average diameter of human hair is about 60 000 to 100 000 nanometres.
- A sheet of paper is about 100 000 nanometres thick.
- Ten hydrogen atoms lined up is equal to one nanometre.

Nanoscience and nanotechnology are multidisciplinary scientific fields that deal with the preparation, processing and application of materials, devices, and systems at the nanometre scale. These fields have attracted great interest among scientists and researchers, which has led to many developments that are used in our daily life.[2]

Nanomaterials are materials that have at least one dimension in the nanometre scale. Scientists discovered that nanoscale-sized materials demonstrate different physical and chemical properties compared to their larger counterparts. For example, gold (Au) displays known, familiar properties of being: (i) a gold-colored, soft, and shiny metal; (ii) a good conductor of heat and electricity; (iii) chemically inert (making it widely used in the jewellery industry); and (iv) its melting point could reach 1064 °C.[63] These properties are present in gold, in dimensions larger than the nanometre range. Reducing the size of gold to the nanometre scale causes drastic changes in its characteristic properties. For example, nanometre-sized gold is not characterized as gold in color. In the nanometre size range, it also becomes a semiconductor, good catalyst, and its melting point reduces to 300 °C (Table 1.1). Many materials at the nanoscale exhibit unique optical, electrical, magnetic, and chemical properties which their larger counterparts do not possess. These unique nanoscale properties have a great impact in technology and medicine. Nanomaterials have become basic components of everyday, modern technologies, and major players in twenty-first century technological advances.[3,4]

1.2 Why Do Properties of Materials Change at the Nanoscale?

Two main effects are responsible for the differences of material properties when sized in nanoscale.

1.2.1 Surface Area–Volume Ratio

Physical properties and kinetics of reactions are strongly related to the number of particles that are exposed to the surrounding environment. This

Table 1.1 Physical and chemical properties of gold at the macroscale and nanoscale.

 Gold (regular size)	 Small pieces of gold at the nanometre scale
Always gold in color	Never gold in color
Soft metal	Not a metal but semiconductor
Chemically inert	Could kill cancer cells
Melting point 1064 °C	Melting point 300 °C

means that the material surface area is the central feature which affects reactivity. Subdividing materials into smaller parts maintains an item's volumetric size but its collective surface area increases. For example, a cube with dimensions of 1 cm × 1 cm × 1 cm has a volume of 1 cm^3, and the area of its six surfaces equals 6 cm^2. By dividing the cube into one million smaller cubes of 0.1 mm × 0.1 mm × 0.1 mm, the whole cube maintains its total volume, but simple calculations reveal that after this division, the collective total surface area of the cubes becomes 100-fold larger than that of the original cube. If the cube is divided further into smaller, non-visible cubes with dimensions of 1 nm × 1 nm × 1 nm, its surface area increases by 10 million (see Sections 6.2.1, 6.2.4, Chapter 6, and 5.2.1.1, 5.2.2.2, 5.3.2, Chapter 5).

Nanoscale-sized materials possess greater surface areas than the same amount of their larger-sized counterparts of the same material (Figure 1.1). This is because the number of atoms and molecules that were previously inside the materials are exposed and become surfaces. This presents a particularly unique and interesting phenomenon at the nanoscale: nanomaterials with extremely high surface area to volume ratios will interact with the surrounding environment more intensively. The large surface areas of nanomaterials influence innate properties, such as the melting point, density, solubility, mechanical properties, electrical conductivity, surface tension, catalytic activity, and spectroscopic properties.[5–11]

1.2.2 Quantum Confinement Effect

The quantum confinement effect is observed when a material size affects its electronic and optical properties. That is, when a semiconductor material absorbs a photon with an energy equal to or higher than its bandgap, it promotes an electron from the valence to the conduction band. This event leaves a hole in the valence band and an exciton is then formed. An exciton is an electron–hole pair associated with coulomb electrostatic interactions between the hole and the promoted electron. According to the rules of quantum mechanics, the distance between the hole and the promoted electron is the Bohr exciton radius. A typical semiconductor material is

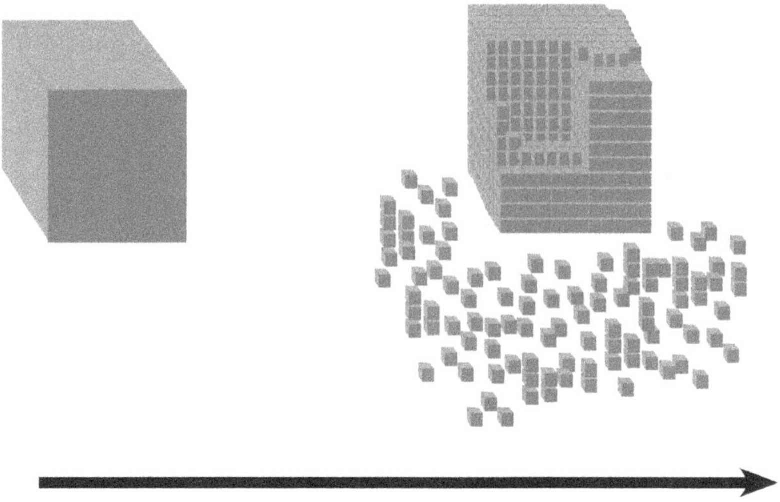

Figure 1.1 Total surface area increases while cutting cube to smaller cubes.

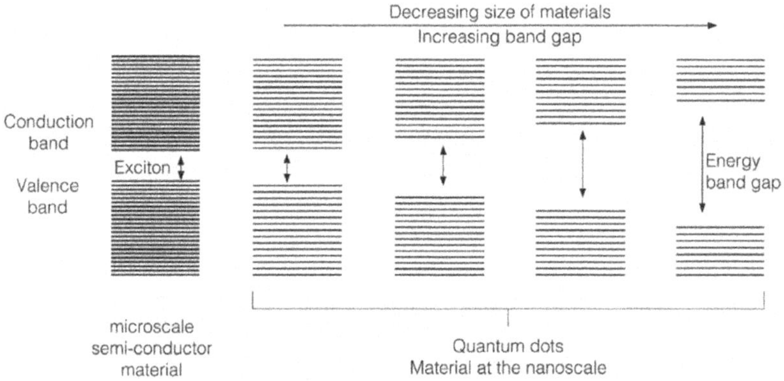

Figure 1.2 Schematic illustration of quantum confinement effect.

larger than the radius of the Bohr exciton when the electron motion is not confined or restricted. On the other hand, when reducing the size of a material to the nanoscale, the material size will lessen to be smaller than the radius of the bulk Bohr exciton. This leads to confinement of electron motion and a congested arrangement of electron holes, resulting in quantum confinement of the electron hole pairs. Consequently, the bandgap increases, and discrete energy levels appear. This increased bandgap creates differing optical and electrical properties. As a result, nanoscale-sized materials could appear to be different colors not normally associated with larger-sized materials of the same type (Figure 1.2).[5,12–17]

1.3 Types of Nanostructured Materials

The most common method to classify nanosized materials is based on the number of dimensions that are not at the nanoscale range. Accordingly, four categories of nanomaterials are classified as the following:

(i) Zero-dimensional (0D) nanomaterials range from 1 to 100 nm in each spatial dimension. That is, electron confinement that occurs in all *x*, *y*, and *z* directions.
(ii) One-dimensional (1D) structures are nanoscale in two dimensions, and one dimension is on the macroscale. Electron confinement occurs in two dimensions, and in one dimension the electrons are free to move along it.
(iii) Two-dimensional (2D) structures consist of one dimension confined to the nanoscale and two dimensions in the macroscale. Electron confinement exists in one dimension, while in the other two macroscale dimensions the electrons are free to move.
(iv) Three-dimensional (3D) structures do not have dimensions in nanoscale. Therefore, they do not have electron confinement in the nano range and electrons move freely along all three *x*, *y*, and *z* directions. Figure 1.3 illustrates the four types of nanostructured materials according to their number of nanoscale dimensions.[18]

Each category of nanomaterials consists of differing types of nanostructures which are categorized according to their shape, morphology, characteristics, and application. The most common representation of 0D nanomaterials are nanoparticles and quantum dots (QD). Nanoparticles are defined as small nanostructured materials with sizes ranging from 1 to 100 nm. The properties of nanoparticles are directly related to their size. Nanomaterial properties change when their size approaches the nanoscale as this increases the ratio of atoms on their surfaces.

Metal nanoparticles, such as gold and silver particles, are distinguished from other nanomaterials by their unique optical properties that arise from a phenomenon called localized surface plasmon resonance (LSPR). LSPR refers

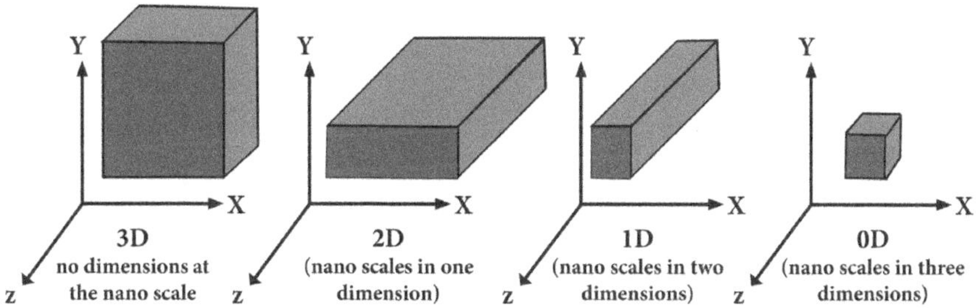

Figure 1.3 Types of nanostructured materials based on number of nanoscale dimensions.

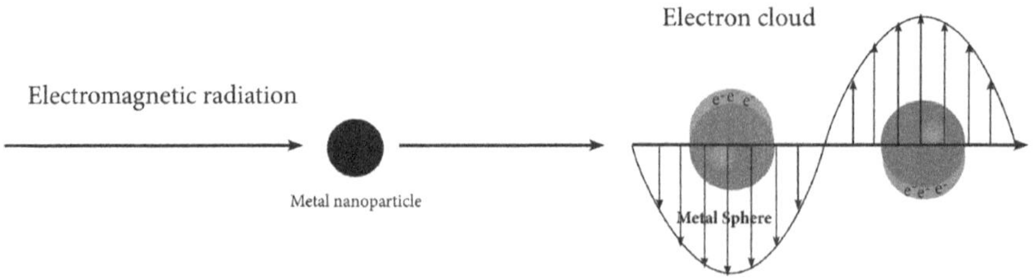

Figure 1.4 Schematic illustration of localized surface plasmon resonance (LSPR) of metal Nanoparticles.

to when metal nanoparticles are irradiated with electromagnetic radiation and the wavelength of the light becomes much longer than the size of the nanoparticles, inducing oscillation of conduction electrons when interacting with the light. This oscillation around the surface of the nanoparticles causes a dipole to oscillate along the direction of the electric field of light. The amplitude of the oscillation reaches its maximum at a specific frequency, thereby denoted as LSPR. The LSPR induces strong absorption of incident light, which can be measured with ultraviolet–visible absorption spectroscopy (UV–vis). This phenomenon is responsible for the unique, unfamiliar colors of the different metallic nanoparticles, illustrated in Figure 1.4.

The LSPR phenomenon is affected by the size and shape of metal nanoparticles. For example, gold nanoparticles have fascinated scientists because of their ability to display a variety of colors, due to the collective oscillation of the electrons at the surface of the nanoparticles, as previously described earlier in this section. Optical absorption is also strongly dependent on the size of gold nanoparticles. Consequently, gold nanoparticles can appear red, orange, purple, and even blue. To produce glossy sheens, nineteenth century artists exploited the colorful range of metallic nanoparticles by applying them onto vessel surfaces or producing stained glass.

Quantum dots are nanostructured particles made of semiconductor materials. In comparison to nanoparticles, the size of quantum dots ranges from 1 to 10 nm, less than the size of nanoparticles. Quantum dots are strongly characterized by the electron confinement effect and this is reflected by their unique array of colors.[19]

One of the molecular allotropes of carbon is fullerene C_{60}. Fullerene exists as a 0D nanoparticle consisting of sp^2 hybrid carbon atoms arranged in five- and six-membered rings, and is widely used in electronics, biology, and medicine.

Nanowires and nanotubes are 1D nanostructures and are characterized by diameters between 1 and 100 nm. Their lengths are much larger than their diameter range. A nanowire is defined as any solid material in the form of a wire with the diameter in the nanoscale range. The ratio between the length of nanowire and its nanoscale diameter is called the aspect ratio. Typical nanowires have aspect ratios of 1000 or more, which signifies that its length is at least 1000 times longer that its diameter. Differing types of nanowires include metallic, semiconductor, and molecular nanowires. All nanowire types are composed of repeating molecular units of organic or inorganic materials. Two nanowire dimensions are affected by quantum confinement due to their sizes being in nanoscale. The length of nanowire is not on the nanoscale and therefore the quantum confinement effect does not take place.

Nanotubes are hollow nanowires with a cylindrical shape. The most well-known nanotubes are called single-walled carbon nanotubes (SWCNT) and they consist of rolled up graphene sheets. Graphene is one of the allotropic structures of carbon. Graphene is a single layer of graphite that is made up of carbon atoms organized into hexagonal shapes. Each graphene carbon atom bonds to three other carbon atoms and leaves one electron in free motion, thus causing high electrical conductivity.

Carbon nanotubes are cylindrical structures made up of graphene sheets with diameters in the nanoscale and lengths in the macroscale. Carbon nanotubes are known for their strength and rigidity, stemming from the strong covalent bonds formed between their carbon atoms. Another form of carbon nanotube is the multiple-walled carbon nanotube (MWCNT), composed of multiple layers of graphene. Each nanotube in a MWCNT is considered a single molecule consisting of millions of carbon atoms. Carbon nanotubes have tremendous electrical, thermal, and mechanical properties that make them attractive candidates in the development of novel materials and applications, including biomedical applications, biosensors, and energy storage. A schematic representation of nanowires and nanotubes is shown in Figure 1.5a and b respectively.[20]

Thin layer structures or nanosheets are typical examples of 2D nanostructures, as the thickness of a nanosheet is in nanoscale. The most well-known nanosheet is graphene. As shown in Figure 1.6, graphene sheets are composed of carbon atoms where each atom is bonded to three other ones and packed into a sequence of hexagonal shapes. On the other hand, single-walled carbon nanotubes and multiple-walled carbon nanotubes are 1D nanostructures made of cylindrical structures of graphene sheets.[21]

Three-dimensional nanostructures are not confined to the nanometre scale in any dimension. This means all three dimensions are at the

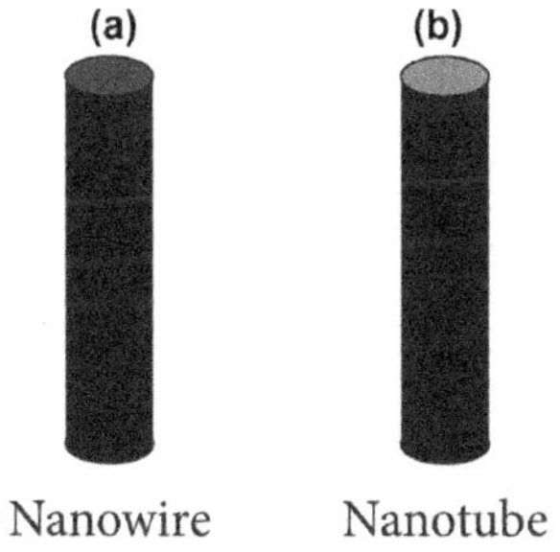

Figure 1.5 Schematic description of: 1.5a, nanowire and 1.5b, nanotube.

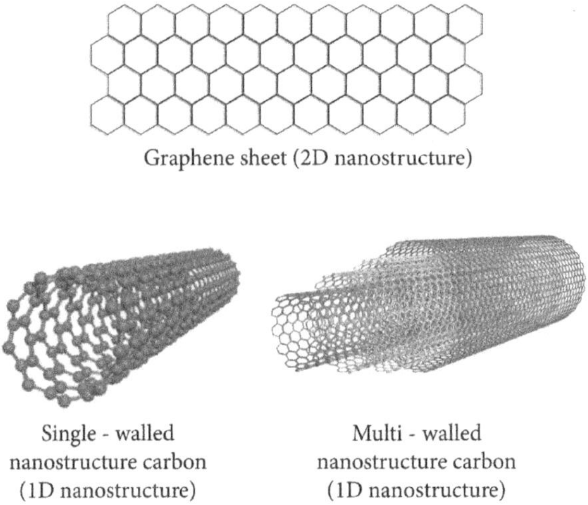

Figure 1.6 An illustration of graphene sheet structure, single-walled carbon nanotube, and multiple-walled carbon nanotube.

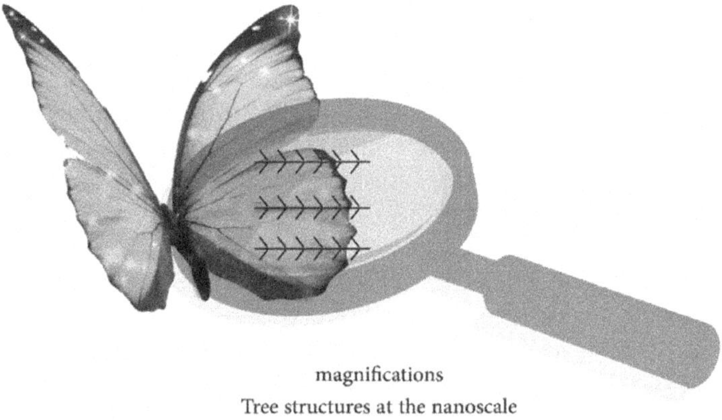

Figure 1.7 Nanostructures of Morpho butterfly wings.

macroscale and the electrons are free to move in all three dimensions. Three-dimensional nanostructures can be composed of dispersed nanoparticles, nanowires, and nanotubes.[22] In nature, many examples of 3D nanostructures exist, including the blue, iridescent-colored wings of the morpho butterfly (*Morpho achilles*). It was discovered that the unique, blue-colored wings of the morpho butterfly are composed of multilayered 3D nanostructures, as illustrated in Figure 1.7.[64]

1.4 History of Nanomaterials

Recent studies have shown that the use of nanostructured materials began in antiquity, without intent or awareness of the ideas of nanoscience and

Figure 1.8 An image in which ancient Egyptian blue was used.

nanotechnology.[23] Five thousand years ago, Egyptian civilization was known to produce a unique, bright blue dye called Egyptian blue (Figure 1.8). In that period, the dye was used frequently throughout the Middle East for painting, statues, tombs, and more. This dye was characterized by its unparalleled blue color that resulted from copper being one of its main components. It ranged in color from light to dark blue, depending upon its composition. Recent studies have proven that this unique blue dye is composed of a mixture of Cuprorivaite ($CaCuSi_4O_{10}$), or calcium copper tetrasilicate, and silicon dioxide (SiO_2) nanoparticles. These compounds are heated to high temperatures ranging between 800–900 °C in order to produce they dye's characteristic range of blue colors. After the end of the Roman era, the Egyptian blue dye , its components and method for creation were lost and ended up in limbo.[24–26]

The production of colored stain glass was a notable trend in Europe the years of 1200–1000 BC, the late Bronze Age. A red glass discovered in the Italian locale of Frattesina di Rovigo is found to be composed of copper nanoparticles. Similarly, red Celtic enamel from 400–100 BC is also found to be composed of copper nanoparticles.[27]

In the fourth or fifth century AD, Romans used gold and silver nanoparticles to create the famous Lycurgus Cup. The cup appears to be one of two different colors, depending on the direction of light passing through it. When light passes from the front, the cup appears green. When light passes through from behind, the cup appears red. Studies attributed this dichroism (two colors) phenomenon to the unique effects produced by the cup's gold and silver nanoparticles.[28]

The Damascus Sword was produced some time between the eighth and seventeenth centuries and was characterized by its unique combination of flexibility and strength. Although swords are known to be made of steel, the rigidity of the steel depends on its carbon and iron concentrations and ratios.

Figure 1.9 An image of the ancient Damascus sword.

The French and Russians attempted to duplicate the strong and flexible features of the Damascus Sword in the early eighteenth century. However, their attempts were proven unsuccessful, and they were unable to discover the secret behind its production. In 2006, scientists were able to reveal the production secret behind the sword's unique features. The Damascus Sword, which is presented in Figure 1.9, contained carbon nanotubes, one of the most durable and flexible materials known to this day.[29]

As discussed above in Section 1.3, metal nanoparticles are characterized by their distinctive colors that differ from their familiar color when they are at the macroscale due to the phenomenon of LSPR. In Europe during the Middle Ages, in order to color medieval stained glass, glass windows in churches were painted with distinctive metal nanoparticle colors (see Figure 1.10). It was discovered in later years that these colors consisted of solutions of nanoparticles of silver, gold and copper.[30,31]

In 1857, nanoscience and nanotechnologies were first introduced to the scientific arena by an English scientist named Michael Faraday. He developed the first scientific procedure to produce the ruby color of gold nanoparticles dispersed in an aqueous solution. In 1908, the scientist Mie explained that the appearance of different metal colors at the nanoscale was due to the theory of quantum size effects.[32] In 1940, silicon dioxide SiO_2 nanoparticles were first produced and used as additives for black carbon in rubber reinforcement.[33]

Later, in 1959, Richard Feynman famously lectured at the meeting of the American Physical Society under the title of: "There is Plenty of Room at the Bottom".[1] Here he postulated, "what would scientists be able to do if they could control the material at the atom and molecular scale and re-arrange it as they want?". He also described a new field of science dealing with individual atoms and molecules to make materials and machines with distinctive properties. This was the first known mention of a new field later termed nanotechnology.[1]

The year 1974 was the first time the term "nanotechnology" was used to describe different methods to produce very small particles.[34] In 1976, the

Figure 1.10 An image of medieval colored stained glass.

Palestinian–American particle physicist Munir Nayfeh renewed a laser technique to detect and measure individual atoms with the highest levels of accuracy and control. He succeeded in detecting one atom out of millions and could revealed its identity for the first time in history; his method was based on excitation of atoms by laser beams.[35] Based on Munir Nayfeh's discovery, the Swiss researchers Gerd Ping and Henrik Rohr invented the Scanning Tunnelling Microscope in 1981.[36] For the first time in history, this microscope allowed scientists to directly image and manipulate atoms and molecules in order to create nanoparticles. In 1991, the Japanese researcher Sumio Lijima invented carbon nanotubes. With their unique mechanical and electronic properties, the creation of carbon nanotubes enabled the manufacturing of distinctive nanostructures.[37]

1.5 From the Optical to Electron Microscope: Seeing at the Nanoscale

Many scientific applications in biology, chemistry, and physics require observing and viewing extremely small sample sizes that are invisible to the naked eye. In the late sixteenth century, optical microscopy was invented as a tool for magnifying and observing very small objects. An optical microscope's technology is based on the use of visible light rays and different lens types in order to magnify a sample. The magnification process depends on shining a light directly onto the sample in a two-lens system. From there, the optical microscope is able to enlarge objects up to a maximum of 2000 times.

This assists the user in being able to view microorganisms and cells in the micrometre size range. However, the limited magnification power of the light microscope was not ideal for scientists and researchers whose research required viewing and analyzing objects in size ranges smaller than the microscale.[38]

In 1931 the electron microscope was invented by German engineers Ernst Ruska and Max Knoll. The principal adaptation to microscopic technology used in the electron microscope was to use rays of electrons instead of visible light. Wave properties of electrons were discovered in a cathode ray experiment performed by J. J. Thomson in 1897. Einstein first proposed that light consists of discrete units of energy known as photons: this dual nature of light became known as the wave–particle duality. Louis de Broglie extended the wave–particle duality principle to electrons. He argued that electrons have both particle and wave features and proposed an equation that expresses the wavelength of an electron beam as:[39]

$$\lambda = h/p \tag{1.1}$$

where λ is the de Broglie wavelength of electron, h is the Planck's constant, and p is the electron momentum. When comparing the wavelength of electron waves to the wavelength of a visible light beam that is used in optical microscopes, the wavelength associated with the electron is much shorter than the wavelength of visible light. The average wavelength of visible light is equivalent to 5000 Å, while the wavelength of electron rays only reaches 1 Å. The discovery of the wave property of electrons is considered a new starting point for the invention of microscopes that rely on electron waves, instead of using visible light rays. This led to much larger magnification power, due to the very short wavelength of electron waves. The magnification power of electron microscopes can reach up to 2.5 million times enhanced magnification of a sample, which enables viewing at the atomic or molecular level. The invention of electron microscopes created an important and useful tool for studying and analyzing very small particles in the nanoscale. Mentioned below are some of the electron microscope types that are widely used in studies of nanoscale materials and structures.[39,40]

1.5.1 Transmission Electron Microscope

The development of magnification devices which rely on the use of electron beams instead of visible light rays have had a great impact on the development of nanoscience and nanotechnology. A transmission electronic microscope (TEM) is one of the most important magnification devices used in this field. It is a form of microscope that magnifies *via* the interaction of electron beams passing through a sample. From there, an image can be obtained on a fluorescent screen or by using a charged-coupled device (CCD) camera. The ability of a TEM to magnify with high accuracy is due to the short wavelength of electrons. This made the device a very important analytical tool used in nanomaterial research. The first TEM was built in 1938

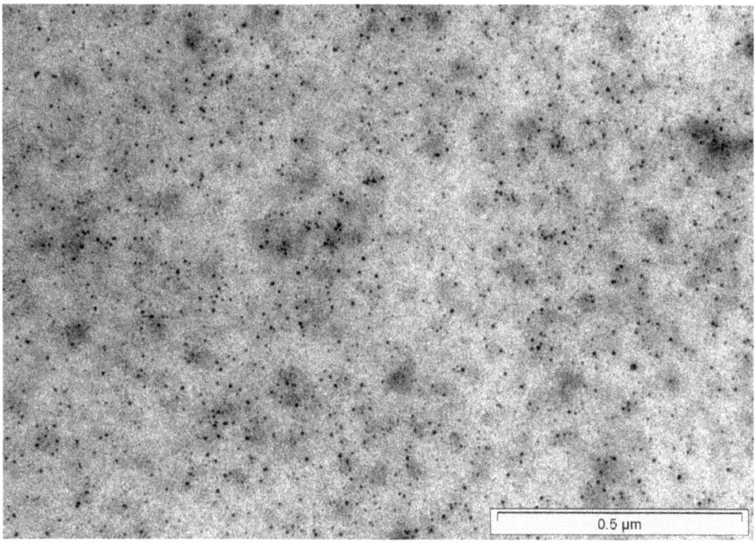

Figure 1.11 An image of magnetite (Fe_3O_4) nanoparticles obtained by transmission electron microscope (TEM).

and contained an internal vacuum system and electromagnetic lenses. The TEM vacuum system enabled electron rays to reach the sample without colliding with gas molecules while the electromagnetic lenses made of magnets directed the electron path.[40] Figure 1.11 presents an image of Fe_3O_4 nanoparticles obtained by a TEM.

1.5.2 Scanning Electron Microscope

A scanning electron microscope (SEM) is a device which can obtain three-dimensional images for analyzing morphology and shape. The SEM shines electron rays onto a sample and a three-dimensional scan of the sample is obtained *via* mapping detailed information from the sample cavities, scratches, and cracks. A magnetic field is used to control the movement of the electron beam over the sample. The electron beam carries large amounts of kinetic energy, but this energy disperses as a variety of signals generated by electron-sample interactions when the electrons are slowed in the solid sample.[40]

1.6 Nanoparticle Behavior in Colloid Solutions

Chemically, substances are divided into two main categories: (i) pure substances, which includes elements and compounds; and (ii) mixtures. Mixtures can be homogeneous or heterogeneous. In general, solutions which are defined as homogeneous mixtures are uniform in composition. Figure 1.12 presents three solutions that seemingly look the same. Solution

Figure 1.12 Three apparently similarly colored solutions of: 1.12a, aqueous solution of red food dye, 1.12b, aqueous solution of gold nanoparticles, 1.12c, an oil in water emulsion.

"a" is a solution of a red food dye dissolved in water. Solution "b" is a solution of gold nanoparticles dispersed in water. Solution "c" is an emulsion of oil droplets suspended in the water phase when the water is colored red. At first glance, all these solutions appear similarly red in color. However, scientifically, these solutions are categorized into three different types: true solutions, colloid solutions, and suspensions. The main difference between these three types of solutions is attributed to the size of dissolved substances.

A true solution is a homogenous mixture composed of one or more solutes dissolved in a solvent. The size of solute particles in a true solution is below 1 nm. A suspension is a heterogenous solution where the size of solute particles ranges from 10 to 1000 μm. In a suspension, the solute particles are large enough that gravity forces cause them to settle downwards unless the mixture is constantly stirred or shaken. Solutions of nanoscale materials are considered colloid solutions. A colloid solution is a heterogenous composition of two phases, a dispersed phase and a dispersion medium. The dispersed phase is composed of nanoscale particles that are evenly dispersed throughout the dispersion medium. The medium can be a solid, liquid, or a gas. The size of dispersed phase particles in colloid solutions ranges from 1 to 10 nm. The word "colloid" is of Greek origin meaning "glue" and was first mentioned by Thomas Graham in 1861. Graham created a "pseudo solution" with a very small diffusion rate. This phenomenon indicated to him that the reason could be due to the presence of "huge" particles in the solution.[41]

The main difference between the three types of solutions is attributed to the size of solute particles. Colloid solutions are composed of nanoscale solute particle sizes between those found in the true solutions and suspensions. Figure 1.13 illustrates the types of solutions categorized according to their solute particle size ranges.[42,43]

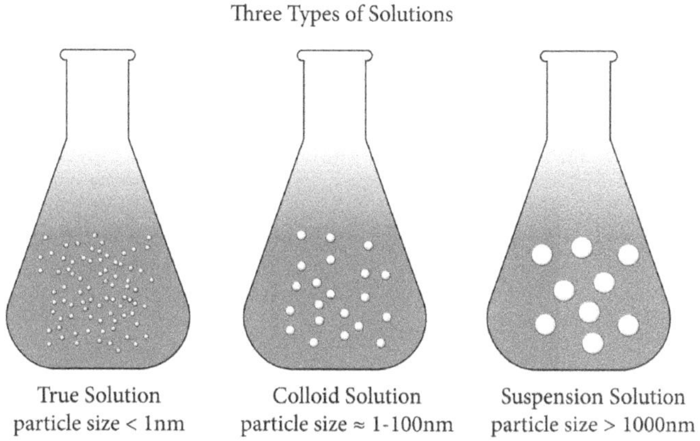

Figure 1.13 Schematic illustration of three kinds of solutions according to solute particle size.

1.6.1 Types of Colloid Solutions

Colloid solutions are defined as heterogenous mixtures of intermediate sized "solute" particles between 1 to 100 nm that do not settle out. Colloid solutions are composed of two phases – a dispersion medium and a dispersed phase of nanoscale particles. Each component of the colloidal solution can exist in three states of matter: gas, liquid and solid. Consequently, different combinations of dispersion mediums and dispersed phases of different states lead to several types of colloidal solutions. For example, sol is the term used when referring to solid nanoparticles dispersed in liquid dispersion medium. If the dispersion medium is an aqueous solution, it is called hydrosol. If the dispersion medium is an organic solvent, the nanoparticle solution is considered an organosol. When nanoparticles are in the gas state and are dispersed in a solid or liquid dispersion medium, it is called a foam. An emulsion is produced by the dispersion of liquid nanostructures in a liquid dispersion medium. An aerosol is the term used when liquid or solid nanoparticles are dispersed in a gas phase. Table 1.2 presents several types of possible colloid solutions, depending on the physical state of dispersion medium and dispersed phase.

Colloidal solutions are differentiated from other types of solutions due to Brownian motion and Tyndall effect properties.[44,45] Brownian motion is a random movement or zigzag-like motion of nanoparticles in colloidal solutions that results from collisions between the nanoparticles and the molecules of the dispersion medium. Figure 1.14 illustrates the random Brownian motion of a nanoparticle in solution.

The Tyndall effect is an optical property that arises from passing a beam of light through colloidal solutions. When a light beam passes through a colloid solution, the light collides with the nanoparticles inside, and this results in light scattering in all directions. This phenomenon is represented by seeing a beam of light inside the solution. The light scattering is attributed

Table 1.2 Types of colloid solutions depending on the physical state of both the dispersion medium and the dispersed phase.

Physical state of dispersion medium	Physical state of dispersed phase	Type of colloid	Examples
Liquid	Solid	Sol	Aqueous colloid solution of gold nanoparticles, paints
Liquid	Gas	Foam	Soap-lather
Liquid	Liquid	Emulsion	Milk, oil in water emulsion
Gas	Liquid	Aerosol	Fog, cloud, mist
Gas	Solid	Aerosol	Smoke
Solid	Liquid	Gel	Jellies
Solid	Gas	Solid sol	Foam rubber
Solid	Solid	Solid sol	Some colored glasses

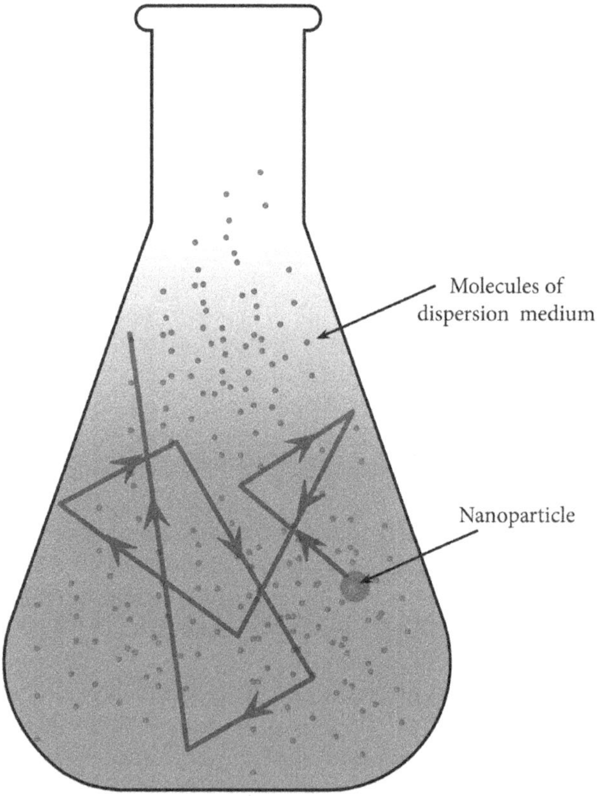

Figure 1.14 Schematic illustration of random Brownian motion of nanoparticles in colloid Solution.

to the size of particles in the solution. In comparison, true solutions have solute particles with sizes below 1 nm. This size is too small to lead to light scattering, and therefore no beam of light is detected. Figure 1.15 illustrates these two solutions, a true solution of silver ions and a colloid solution of

Figure 1.15 Description of the Tyndall effect: passing of laser beam through, a, colloid solution of silver nanoparticles, b, true solution of silver ions.

silver nanoparticles. In image 1.15a, a shiny line inside the colloid solution is visible after a laser beam passes through it. In image 1.15b, nothing is detected when a laser beam passes through the true solution.

1.6.2 Stability of Colloidal Solutions

Technological and medical applications of materials at the nanoscale are mainly based on using stable colloidal solutions. Random Brownian motion of nanomaterials in colloidal solutions increases the possibility of collision between the nanoparticles. Colloid stability depends on the interactions between them when they collide. Van der Waals forces are considered a main attractive force between nanoparticles in solutions.[46]

There has been increasing attention put upon the particle–particle interactions in nanomaterials due to their large surface area. If attractive forces between nanoparticles exceed repulsive forces, then the nanoparticles tend to cluster together and aggregate into larger-sized particles that eventually precipitate under the influence of gravitational forces. The aggregation process of nanoparticles is represented by several stages as follows and depicted in Figure 1.16.[46]

1. Flocculation, where the aggregation process is reversible
2. Coagulation, where the formation of large, aggregated particles becomes an irreversible process
3. Sedimentation, where the aggregated nanoparticle precipitation is caused by the force of gravity

To prevent the process of aggregation between the nanoparticles and to obtain a stable colloidal solution, the repulsive forces must overcome the forces of attraction between them. The main attraction forces between nanoparticles in colloidal solutions are van der Waals forces. These forces

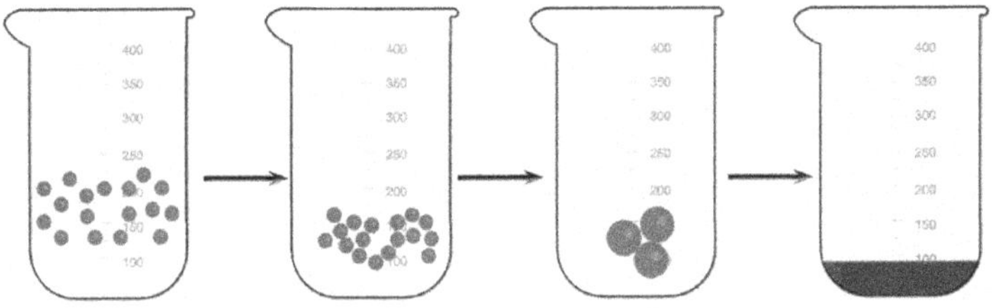

Figure 1.16 Schematic illustration of aggregation process of unstable nanoparticles.

have great effects on colloid stability and can lead to aggregation, which destroys the structure of the colloid solution. In order to prevent nanoparticle aggregation, three stabilization methods exist and are described in the following Section 1.6.2.1.[47]

1.6.2.1 *Electrostatic Stabilization*

The most effective way to balance the attractive forces between nanoparticles in solution is to create electrical repulsion forces between them. This is done by dispersing positive or negative charges onto the nanoparticle surface. Surface electrical charges could be created by one or more of the following mechanisms:

- Ionization of functional groups – examples include organic molecules in the formulas of RCOOH, which ionize to $RCOO^-$ in aqueous solutions; amines, RNH_2, which ionize to RNH_3^+; and SiOH, which ionizes to SiO^- in aqueous solutions. This mechanism depends on the chemical structure of the nanoparticles.
- Physical or chemical adsorption of electrically charged materials applied to the nanoparticle surface – there are different types of electrically charged materials that could adsorb to the surface of nanoparticles including simple ions, polymers, or surfactants.[42]

1.6.2.1.1 Electrical Double Layer. The stabilization of a colloid solution *via* generating electrically charged nanoparticles results in the formation of an electrical double layer. This electrical double layer consists of parallel layers of co-ions and counterions which surround the nanoparticle. The electrical double layer is created mainly from the different attraction and repulsion forces between the ions in the colloidal solution. According to this theory, the presence of electric charges on the colloidal particle surface area affects the dispersion of ions in solution. Counterions are ions with opposite charges that stay close to the particle surface, due to the electrostatic attraction between them. Co-ions are ions with similar electric charges distributed within the solution. The specific electrical charge distributions around the surface of the particles and the effects of Brownian motion cause the formation of the electrical double layer.

An electrical double layer consists of two main regions, the Stern layer and the diffuse layer. The inner region is the Stern layer, which is an electrically charged surface of nanoparticles. The diffuse layer is the outer region, composed of dispersed, oppositely charged ions in solution, which distance themselves from the particle surface. In 1994, the scientist Stern developed a model in which the inner and outer regions of the electrical double layer were separated by an imaginary plane called the Stern plane. The outer part of the double layer was characterized by Gouy in 1910 and Chapman in 1913. They assumed that the dispersion of ions is based on a Boltzmann distribution. Figure 1.17 portrays the structure of an electrical double layer of a negatively charged particle incorporated into a distribution curve that illustrates the change of electrical potential generated as a function of the distance from the surface of the nanoparticle.[46]

Experimentally, the electrical potential of charged colloidal particles can be measured only at a certain distance from its surface. As shown in Figure 1.17, the distance that exists behind the Stern plane and at the edge of the diffuse region indicates the charges associated with the particle's charge

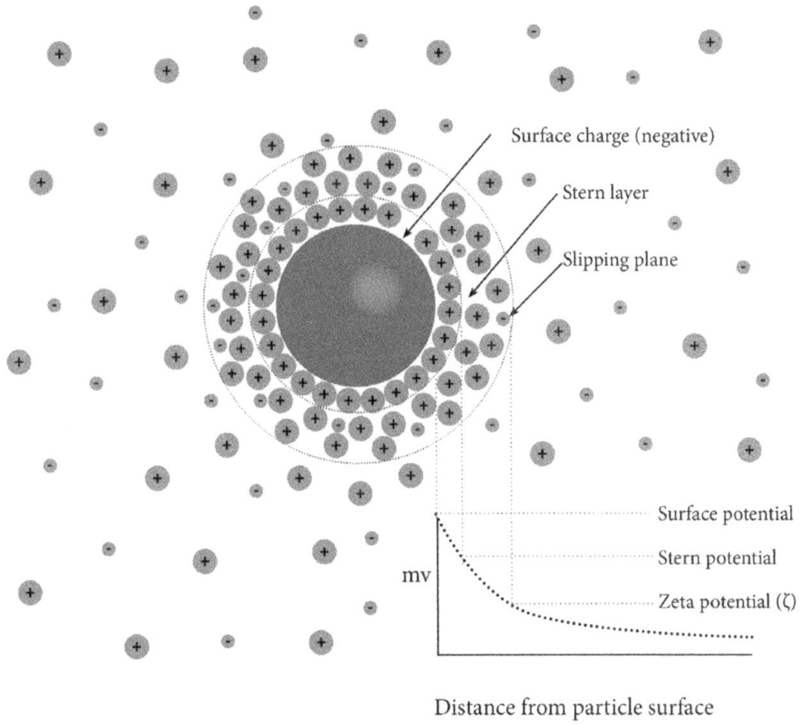

Figure 1.17 Schematic illustration of an electrical double layer of negatively charged nanoparticle. and a curve represents electrical potential as a function of distance from the nanoparticle surface.

Table 1.3 Possible values range of zeta potential and its relationship on the stability and instability of colloidal solutions.

Values of zeta potential/mV	Degree of stability of a colloidal solution
0 to ± 5	Flocculation and coagulation
± 10 to ± 30	Unstable
$> \pm 30$	Stable

and its stability. This phenomenon is called the zeta potential and is represented by the Greek letter zeta (ζ). In principle, the numerical value of the electrical potential decays as one moves away from the particle surface, but the zeta potential can still attest to the surface charge. The measured magnitude of the zeta potential is an indication of the stability of a colloid solution. Lower zeta potential values specify that the electrical charge is not sufficient to prevent attraction between nanoparticles. Higher zeta potential values indicate that the electrostatic repulsion forces between the nanoparticles in colloid solution could exceed the attractive forces. Higher zeta potential values imply that the colloid solution is stable and that the nanoparticles will not aggregate. Table 1.3 presents different possible zeta potential values and their relationship to the stability or instability of the colloidal solution.[46]

A colloid solution stabilized by electrostatic mechanism is greatly affected by the addition of different electrolytes. This is related to the simple interaction between electrically charged species. The addition of electrolytes with electrical charges opposite to the charge of the nanoparticles (counterions) could gradually lead to naturalization. This affects the colloid solution stability and could lead to destroying the stabilization mechanisms and creating nanoparticle aggregation.

Critical Coagulation Concentration (CCC) is defined as the minimum electrolyte concentration required to coagulate colloidal particles. Adding counterions to a stable colloid solution causes the outer part of the electrical double layer around the charged nanoparticle to compress. From this, an energy barrier is created which is expressed by connecting the particles to distances where van der Waals forces will be significant. This results in nanoparticle coagulation which precipitates under the influence of gravity. Figure 1.18 shows how adding counterions into a solution leads to precipitation of the aggregated nanoparticles from a stable dispersion.

This process is accompanied by a gradual decrease of zeta potential value. For example, if the nanoparticles are stabilized by acquiring negative electrical charges, the addition of positively charged ions leads to minimizing its negative charge, and this is expressed by a lower zeta potential value. In most cases this results in complete naturalization, reflected by a zeta potential value of zero. The zeta potential value of zero is called the isoelectric point.[46]

In an aqueous colloidal solution, the pH value affects the zeta potential and the stability of the solution. If we assume a colloidal solution of negatively charged nanoparticles, the addition of an alkali leads to increasing the

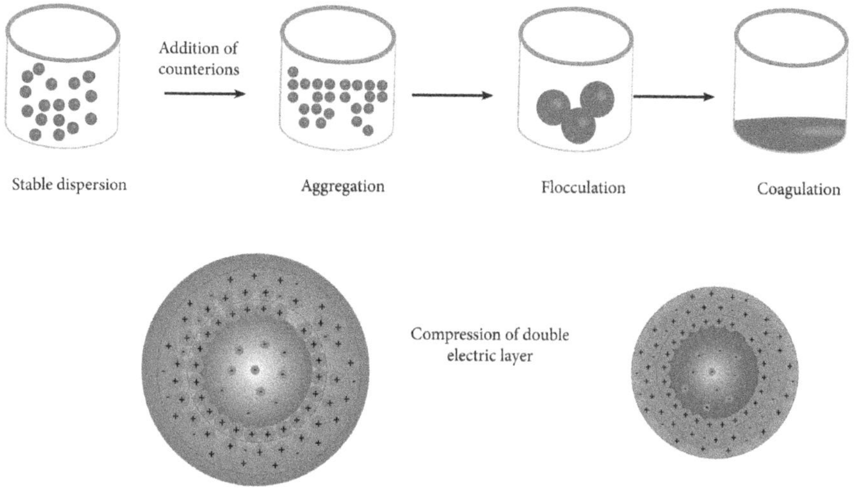

Figure 1.18 Schematic illustration of aggregation process of electrical charged nanoparticles induced by adding counterions.

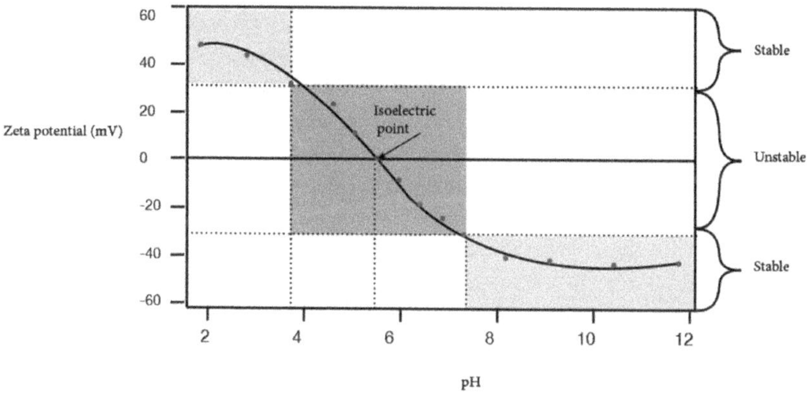

Figure 1.19 A curve represents zeta potential values *versus* pH.

negative charges and this results in higher zeta potential values. If the pH of the solution is lowered by adding an acid, the zeta potential will decrease. The negative charges of the nanoparticles reduce and thereby the solution reaches its isoelectric point. It is very important to note that adding more acidic molecules could lead to positively charged nanoparticles and this restabilizes it by acquiring opposite electrical charges. Figure 1.19 presents the possible curve of the zeta potential value as a function of pH.

1.6.2.1.2 DLVO Theory. In 1940, the scientists Derjaguin, Verwey, Landan, and Overbeek (DLVO) developed a quantitative theory of colloidal stabilization. According to this theory, the stability of particles within a solution depends on the potential energy, V_T, of the particles. This energy results from a balance between competitive attraction and repulsion forces between the nanoparticles that are created by the double electrical layer when:

$$V_T = V_A + V_R + V_S \quad (1.2)$$

where V_A is due to attraction forces and is expressed by:

$$V_A = -A/(12\pi D^2) \quad (1.3)$$

and A is the Hamaker constant that depends on the nature of material and D is the distance between the nanoparticles. V_R is due to repulsive forces represented by the equation below:

$$V_R = 2\pi\varepsilon a \Psi_d{}^2 e^{-KH} \quad (1.4)$$

where ε is the dielectric constant of the dispersion phase, a is the radius of the nanoparticle, Ψ_d is the electric potential on the Stern plane of the electric double layer, $1/k$ is the double electric layer thickness. V_S is the potential energy derived from the solvent. V_S does not significantly contribute to the final balance of the forces existing between the particles.

According to the DLVO theory, the stability of a colloidal solution is a function of the attractive and repulsive forces that are created by the double electric layer when particles collide during Brownian motion. The electrostatic repulsive forces create an energetic barrier that prevents the particles from approaching too closely to one other. However, when the attractive forces between the particles overcome the energetic barrier, aggregation takes place and the colloidal solution is destroyed.[48] Figure 1.20 illustrates the total potential energy *versus* the distance between nanoparticles in solution.

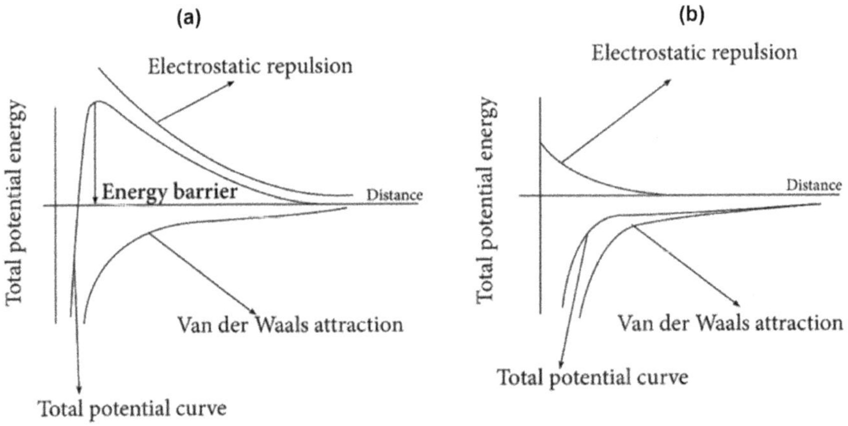

Figure 1.20 A curve that describes the total potential energy between two particles as a function of the distance between them in the case of a, a stable colloid solution, b, unstable colloid solution under aggregation process.

1.6.2.2 Steric Stabilization

Steric stabilization of colloidal solution occurs by the physical or chemical adsorption of macromolecules onto the nanoparticle surface. *Via* this process the macromolecules create a steric barrier against aggregation. Such macromolecules belong to families of polymers, copolymers, amines, and thioethers, as well as solvents such as tetrahydrofuran (THF) and THF/methanol (MeOH), and alcohols with long hydrocarbon chains and surfactants. In general, hydrophobic molecules are used in an organic dispersion medium and the colloid solution is called an organosol. When hydrophilic molecules are used in an aqueous dispersion medium the colloidal solution is called a hydrosol. Figure 1.21 depicts schematic illustration of steric stabilized nanoparticles.[46]

1.6.2.2.1 Mechanism of Steric Repulsion. When there are two nanoparticles with long chains attached to their surface, the interaction between them will be significant as they approach a distance equal to at least twice the thickness of each layer. Quantitatively, the degree of stabilization is described by the energetic change caused because of the interaction between the two layers of the macromolecules around each nanoparticle. The change in free energy that occurs as a result of the interaction between the macromolecules layers is called the Gibbs energy and is defined in the following equation:

$$\Delta G = \Delta H - T\Delta S \tag{1.5}$$

where ΔG is the Gibbs energy, ΔH is enthalpy, T is temperature, and ΔS is entropy. If the free energy is negative due to the interaction, then the nanoparticles will be unstable and undergo aggregation. However, positive free energy causes a stable dispersion of the nanoparticles. In isothermal conditions, the degree of stabilization is a function of the enthalpy and entropy change of the system.

Over the years, several theories have been developed to explain the steric stabilization mechanism. These theories attempt to mathematically

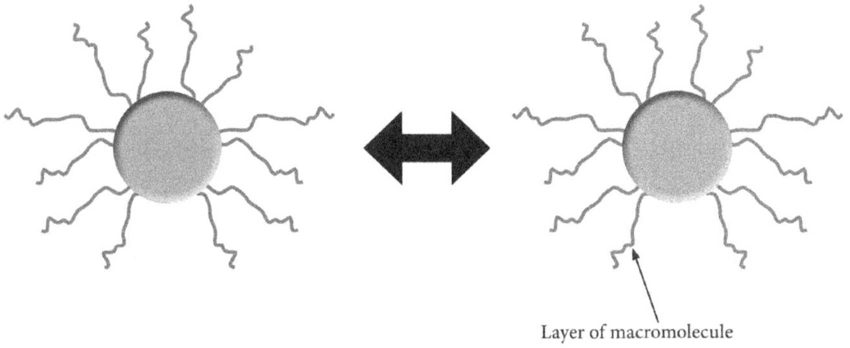

Figure 1.21 Schematic illustration of steric stabilized nanoparticles in colloid solution.

calculate the energetic change due to the overlap between two long chains around each nanoparticle as they approach one another. Accordingly, there is no penetration of one chain layer within the other as two particles approach one other. Thus, when two nanoparticles approach each other at a certain distance, the long chains around their surfaces are compressed and the chain segments lose their entropic configuration. This is when the chain segments have fewer configurations in the compressed state compared to the uncompressed state. This decrease in entropy causes an increase in free energy, repulsion between the nanoparticles, and prevention of aggregation. According to this theory, the enthalpy change between the adsorbed long chains is negligible and therefore it can be assumed that:[46]

$$\Delta G = -T\Delta S \tag{1.6}$$

In general, a polar surface of a nanoparticle is defined as an area with acidic or basic sites. Polymer polar functional groups are usually acidic or basic groups in the molecular structure of the polymer. The adsorption of polymers onto the surface of different particles is a function of the chemical structure of its functional groups and the surface of the particle. For example, it has been found that polyamides and polydimethyl siloxane can adsorb to highly polar nanoparticles but not to particles with low polarity. The scientists Ullman and Ellerstein proved that the presence of highly polar groups embedded in the polymer molecular structure causes adsorption onto the particles. This phenomenon is a result of the interaction of the functional groups with the basic or acidic sites on the particle surface.[49]

In some cases, in addition to the presence of the polymers in the solution, the dispersing medium molecules are very important due to their potential to also adsorb onto the particle surfaces. The competition between the dispersion medium molecules and the polymers is determined by the chemical nature of the particle surfaces. For example, poly(methyl methacrylate) weakly adsorbs onto silica particles when dioxane is the dispersion medium, because this basic solvent "acquires" the acidic surface of the particles and prevents the polymer from adsorbing.

Steric stabilization has several advantages when compared with electrostatic stabilization and are described as follows:

1. Steric stabilization is not affected by the presence of electrolytes in solution.
2. Steric stabilization can be applied to both hydrosols and organosols.
3. In electrostatic stabilization, the flocculation process is due to the addition of counterions, and this process is usually irreversible. The flocculation process of steric stabilized nanoparticles is reversible.

1.6.2.3 Depletion Stabilization

Depletion stabilization is a mechanism which occurs in the presence of long chain polymers in colloidal solution when not adsorbed onto the surface of

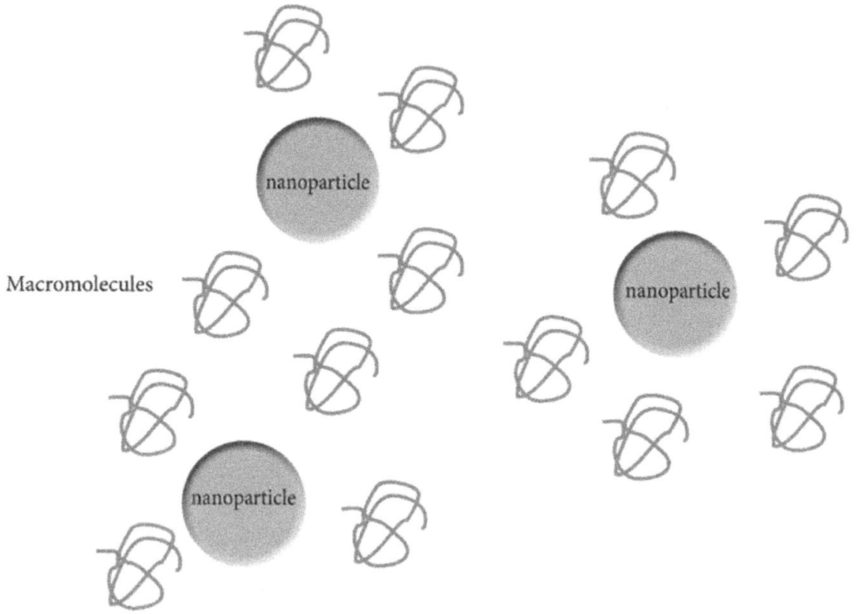

Figure 1.22 Schematic illustration of depletion stabilization of nanoparticles in colloid solution.

the particles, as shown in Figure 1.22. The large dimensions of these chains can compete with van der Waals attractive forces among the nanometre particles. The prevention of particle aggregation is due to the presence of these molecules in solution freely, *i.e.*, there is no interaction between the polymer and the particle.[46]

1.6.3 Ferrofluids

To prepare magnetic materials in liquid state, solid iron must be melted and maintained at a high temperature in order to maintain the magnetic material's liquid state. Nanoscience has revolutionized the production of solutions with magnetic properties without the need to heat materials to high temperatures. These solutions are called ferrofluids. Ferrofluids are colloid solutions of magnetic nanoparticles dispersed in either an aqueous or organic dispersion medium. Creating a stable dispersion of magnetic nanoparticles is considered a significant challenge due to the magnetic attractive forces between the particles and the van der Waals attractive forces. In order to prevent aggregation of the magnetic particles in solution, they must be provided with sufficiently strong enough repulsive forces in order to overcome these strong attractive forces. The stabilization of these solutions is determined by the balance between these strong attractive forces and the repulsive forces caused by steric or electrostatic stabilization. Common solutions of this type are ferrofluids composed of nanoparticles of magnetite (Fe_3O_4), cobalt, and nickel.

Ferrofluids are widely used and in recent decades many chemists and physicists have attempted to synthesize these solutions. Papell was the first scientist to succeed in stabilizing magnetic nanoparticles in 1965. Afterwards, many scientists began to study and synthesize them. The widespread use of ferrofluids is attributed to their unique features as described below.

1. Nanoparticles in solution are strongly attracted towards the direction of an external magnetic field.
2. They absorb electromagnetic energy at a certain frequency and heat up.
3. When exposed to an external magnetic field, their physical properties can change.

These unique features of ferrofluids make them attractive candidates for various applications in the fields of technology, biology, and medicine.[50]

1.6.4 Preparation of Nanomaterials

Over the years various methods of preparing nanomaterials have been developed. In general, these methods have been divided into two main categories, top-down and bottom-up. As illustrated in Figure 1.23, top-down methods are based on breaking down macroscale materials to nanoscale dimensions by applying different energy sources. The bottom-up approach is based on chemical methods where atoms, molecules, and ions are built up to nanoscale dimensions.[42]

1.6.4.1 Top-down Methods

1.6.4.1.1 Mechanical Grinding. Mechanical milling is a method that has attracted great interest due to its simplicity and low cost. This method

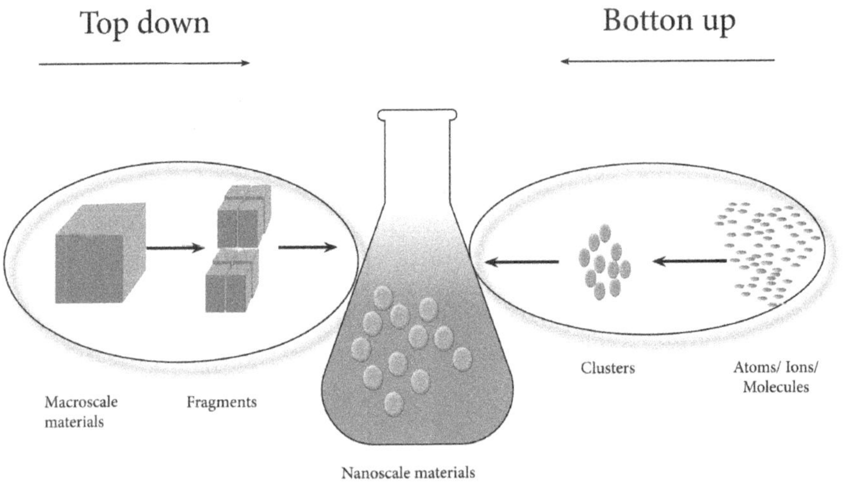

Figure 1.23 Description of bottom-up and top-down approaches for the preparation of nanomaterials.

is based a milling technique where strong mechanical shear forces are used to grind down macroscale bulk materials to particles in nanoscale dimensions. For example, this method is used to fabricate metal nanoparticles ground from high-energy steel ball milling.

1.6.4.1.2 Laser Ablation. Another top-down nanomaterial fabrication approach is based on laser ablation of macroscale bulk metal immersed in liquid solution. The metal is placed at the bottom of a glass vessel inside the solution and laser irradiation is directed towards the metal, thereby causing it to break down into smaller pieces. Color changes of the solution gives an indication that nanoparticle formation is taking place. In order to prevent the aggregation of the metal nanoparticles being produced, different kinds of materials are dissolved into the solution prior to the laser ablation. The materials which prevent nanoparticle aggregation can include polymers or surfactants, in order to achieve steric or electrostatic stabilization as displayed in Figure 1.24.[51]

1.6.4.2 Bottom-up Methods

1.6.4.2.1 Reduction. Reduction is a bottom-up method of creating nanoparticles in a solution by reducing metal ions which are below 1 nm in size. The creation of nanoparticles is based on self-assembly and the clustering together of reduced metal ions into larger sized scales that become nanoscale particles.[42] Oxidation–reduction reactions are based on the transfer of electrons from a reducer material to metal ions in solution

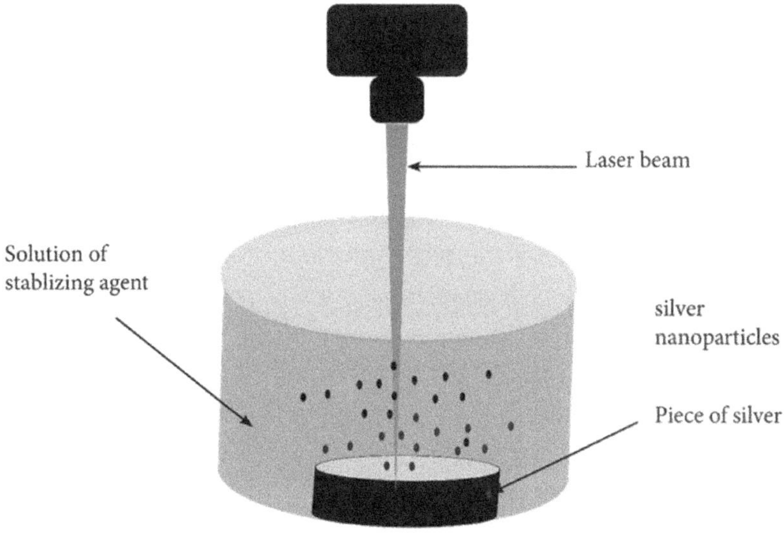

Figure 1.24 Schematic illustration of top-down laser ablation synthesis of metal nanoparticles.

to form zero valent metal atoms. They are considered nucleation seeds for growing metal nanoparticles. The reaction is described as follows:[43]

$$m\mathrm{Me}^{n+} + n\mathrm{Red} \rightarrow m\mathrm{Me}^0 + n\mathrm{O}_x \quad (1.7)$$

The driving force of this reaction is the difference between the oxidation–reduction potentials of the two half reactions, ΔE. ΔE determines the equilibrium constant of the reaction, K_e, according to the Nernst equation:

$$\ln k_e = nF\Delta E/RT \quad (1.8)$$

where F is the Faraday constant, R is the gas constant, and T is temperature. Thermodynamically, the reaction occurs when ΔE is positive, which indicates that the oxidation–reduction potential of the reducer must be more negative than that of the metal ions. This difference must be greater than 0.3–0.4 V. Therefore, electronegative metals with $E_0 < -0.2$ V require the use of strong reducers, and in many cases, they must be applied under specific conditions of pressure and temperature. However, metals with $E_0 > 0.7$ V, such as Au, Rh, Ag, Pd, and Pt, rapidly react with weaker reducers and under normal conditions.

Many metal colloid solutions can be prepared by the reduction of metal ions in solution using different reducers under appropriate conditions. Table 1.4 presents various metal ions, reducers, and appropriate reaction conditions.[43]

The three main steps described below present the process of forming metal nanoparticles. The steps include the reduction of metal ions to zero valent metal atoms, nucleation, and growth of metal nanoparticles.

- Step 1: Reduction – zero valent metal atoms form as a result of an oxidation–reduction reaction and the transfer of electrons from a reducer to metal ions, as described by the formula:

$$m\mathrm{Me}^{n+} + n\mathrm{Red} \rightarrow m\mathrm{Me}^0 + n\mathrm{O}_x \quad (1.9)$$

Table 1.4 Reducing agents, and reaction conditions for the reduction of various metal ions.

Metal ions	Reducer	Reaction conditions	Rate of reaction
Au^{3+}, Pt^{4+}, Pt^{2+}, $Pd^{2+2\backslash}$	Organic acids, alcohols	≥70 °C	Slow
Ag^{+}, Rh^{3+}, Hg^{2+}, Ir^{3+}	Sugars, hydrazine, H_2SO_3, $NaBH_4$	Ambient	Fast
Cu^{2+}, Re^{3+}, Ru^{3+}	Polyols	>120 °C	Slow
	Aldehydes, sugars	70–100 °C	Slow
	$NaBH_4$	Ambient	Fast
Cd^{2+}, Co^{2+}, Ni^{2+}, Fe^{2+}, In^{3+}, Sn^{2+}	Polyols	>180 °C	Fast
	$NaBH_4$	Ambient	Slow
Cr^{3+}, Mn^{2+}, Ta^{5+}	$NaBH_4$	Ambient	Slow
	Radicals	Ambient	Fast

- Step 2: Nucleation – zero valent metal atoms produced from the reduction reaction become insoluble in solution. Therefore, they gradually aggregate to become larger groups called "embryos" as described in the formula:

$$x\mathrm{Me}^0 \leftrightarrow (\mathrm{Me}_x{}^0)_{\mathrm{em}} \tag{1.10}$$

As more zero valent metal atoms are formed in solution, the embryos reach a critical size and separate from the solution to become solid particles {nucleus, $(\mathrm{Me}^0{}_n)_{\mathrm{nucl}}$} and this process is defined as:

$$(\mathrm{Me}^0{}_x)_{\mathrm{em}} + y\mathrm{Me}^0 \rightarrow (\mathrm{Me}^0{}_x)_{\mathrm{nucl}} \tag{1.11}$$

The number of nuclei and their size depend on several parameters: the concentration of metal ions, oxidation–reduction potential of the reaction, temperature, and solvent viscosity.

- Step 3: Growth – As additional metal atoms continue to form in solution, the nucleus continues to grow, and the first particles are produced:

$$(\mathrm{Me}^0{}_n)_{\mathrm{nucl}} + m\mathrm{Me}^0 \rightarrow (\mathrm{Me}^0{}_{n+m})_{\mathrm{pp}} \tag{1.12}$$

These first particles are unstable due to their high free energy. Therefore, they gradually aggregate to become colloidal metallic nanoparticles $(\mathrm{Me}^0{}_{\mathrm{p}})$:

$$(\mathrm{Me}^0)_{\mathrm{pp}} + n\mathrm{Me}^0 \rightarrow (\mathrm{Me}^0)_{\mathrm{p}} \tag{1.13}$$

$$m(\mathrm{Me}^0)_{\mathrm{pp}} \rightarrow (\mathrm{Me}^0)_{\mathrm{p}} \tag{1.14}$$

Figure 1.25 illustrates the mechanism for the formation of colloidal metal particles by the reduction method.

Various studies have demonstrated the possibility of producing metallic nanoparticles by using a variety of plant extracts. These extracts have two main functions: (i) to reduce metal ions, and (ii) to act as stabilizing agents in order to prevent the aggregation of nanoparticles. These reactions are very desirable because of their environmentally friendly properties. Silver and gold nanoparticles have been prepared by this method when their ions were mixed with various plant extracts in aqueous solution.[23]

1.6.4.2.2 Electrochemical Method to Produce Metal Nanoparticles. Electrochemical cells are widely used to produce various types of metal nanoparticles. The idea is based on using an anode made of the same metal to be prepared in the form of nanoparticles immersed in aqueous solution of metal salt. The general process of electrochemical synthesis is described as follows:

$$\text{Anode: } \mathrm{M}_{\mathrm{bulk}} \rightarrow \mathrm{M}^{n+} + ne^- \tag{1.15}$$

$$\text{Cathode: } \mathrm{M}^{n+} + ne^- + \text{Stabilizer} \rightarrow \mathrm{M}_{\mathrm{colloid/stabilizer}} \tag{1.16}$$

$$\text{Sum: } \mathrm{M}_{\mathrm{bulk}} + \text{Stabilizer} \rightarrow \mathrm{M}_{\mathrm{colloid/stabilizer}} \tag{1.17}$$

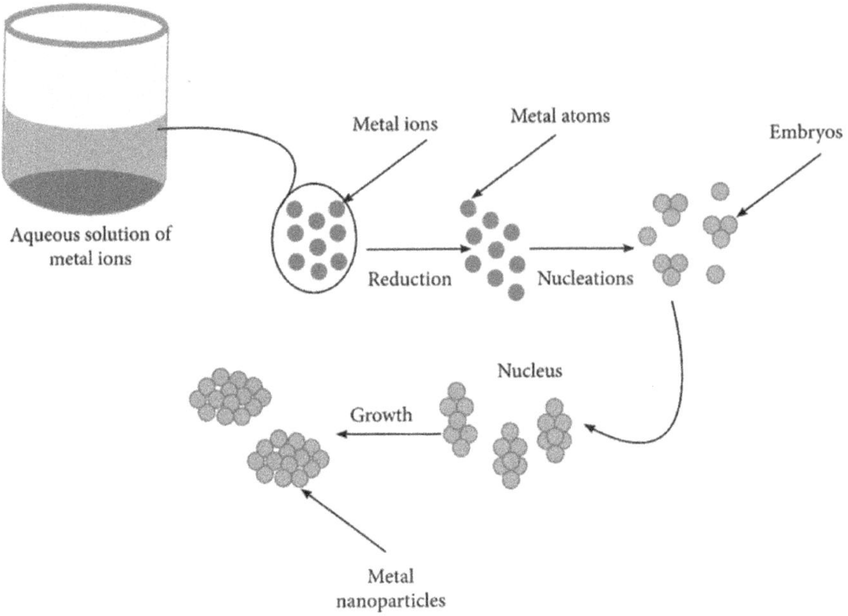

Figure 1.25 Illustration of the mechanism of producing metal nanoparticles by reduction method.

The main processes to form metal nanoparticles in electrochemical cells are described below:

1. Oxidation of metal anode.
2. Direct the flow of metal ions produced in the anode towards the cathode.
3. Reduce metal ions in the cathode to zero valent metal atoms.
4. Primary nuclei form in order to later grow into nanoparticles.
5. Grow nuclei to nanoscale particles.
6. Stabilize nanoparticles to prevent aggregation *via* introducing materials that act as stabilizers.
7. Precipitate metal nanoparticles.

The electrochemical method to produce metal nanoparticles is illustrated in Figure 1.26.

An important advantage of the electrochemical method is the ability to control particle size. This can be done *via* adjusting experimental parameters such as changing the distance between the electrodes along with altering reaction times and temperatures.[42,52]

1.6.4.2.3 Preparation of Nanometals in Gas Phase. The main principle behind the preparation of metal particles in a gas phase is to rapidly cool metal atoms in the gas state using a cold, inert gas. This process involves placing gaseous metal atoms into a furnace of very high temperature,

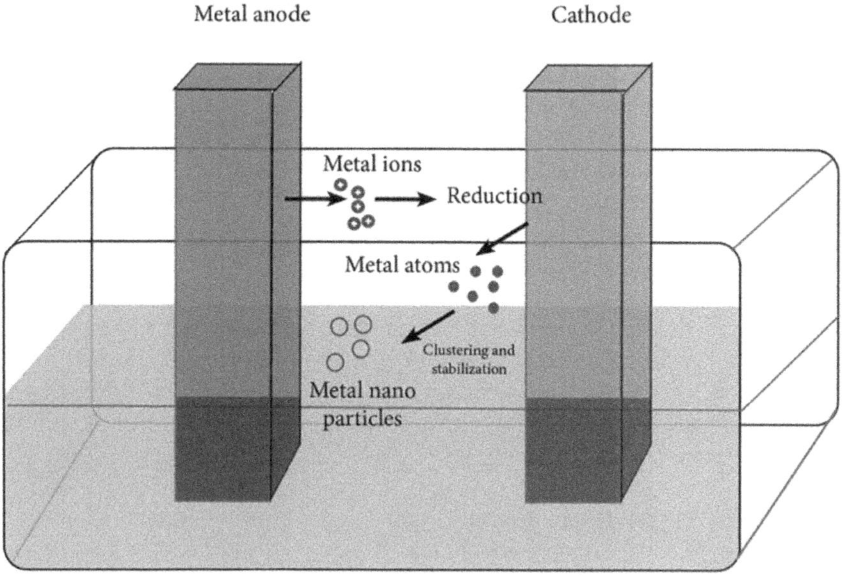

Figure 1.26 An illustration of electrochemical method for the preparation of metal nanoparticles.

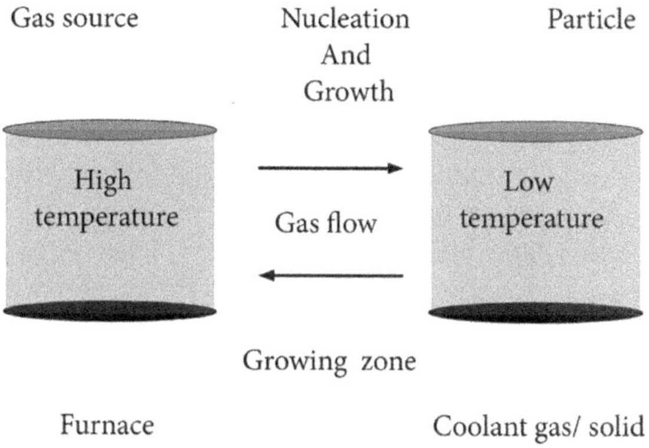

Figure 1.27 Schematic illustration of gas phase preparation of nanoparticles.

within microseconds the atoms move towards high temperature gradient per centimetre space and are cooled by an inert gas. As shown in Figure 1.27, the formation of metal nanoparticles takes place between two ends of extremely different temperatures. The first one exhibits high temperatures whereas the second one is performed under very low temperatures. In this extreme passage many embryos of nuclei are formed. These nuclei embryos aggregate together and establish the nucleation phase. The nuclei continue to grow until they expand beyond the growing zone, an area where there are no more atoms or nuclei. Finally, the nanoparticles formed are obtained at room temperature by cooling in a gas flow.[53]

1.6.4.2.4 Ultrasound for the Preparation of Nanomaterials. Sound is defined as wave vibrations transmitted through an elastic medium. In the middle of nineteenth century, the scientist Galton attempted to measure the hearing threshold of humans and animals. To this end he developed a whistle capable of producing sound waves at different frequencies. Using this whistle, he discovered that human's hearing range is between 16 Hz–18 kHz. Ultrasound is defined as a sound frequency of 20 kHz and higher and is considered beyond human hearing.[54]

Sonochemistry is a field that defines and characterizes chemical reactions under ultrasound irradiation. It has been found that ultrasonic irradiation of liquids leads to the occurrence of chemical reactions that consume high energy. The enormous energy obtained in sonochemistry stems from a physical phenomenon called acoustic cavitation. Acoustic cavitation is the formation of gas bubbles inside a reaction solution. The process of acoustic cavitation includes three steps:

1. The formation of gas bubbles inside an irradiated solution.
2. Growing bubbles to a critical size.
3. Bubble collapse.

According to the hot spot theory, bubble collapse creates extreme conditions of pressure and temperature that breaks chemical bonds. Significant thermal energy is then released which creates localized, high temperature areas called hot spots. When the heating process is adiabatic, there is no transfer or loss of heat to the environment and therefore the hot spots catalyze other chemical reactions. The collapse of gas bubbles is adiabatic and takes place in the very short time span of a nanosecond. Many experiments have shown that upon bubble collapse the temperature can reach up to 5000 K, the pressure up to 1000 atm, and the cooling rate above 10^9 °C min^{-1}. These parameters illustrate the enormous effects ultrasound irradiation can have when applied to a liquid and explain the occurrences of chemical reactions.[55–57]

1.6.4.2.4.1 Sonochemistry in Aqueous Solutions. Sonochemistry is a process based on the creation and collapse of bubbles in a solution. Generally, there are two regions of importance: the region within the bubble and the immediate vicinity surrounding the bubble. The cavitation bubble contains vapor from the solvent or other volatile reagent(s) in the solution. When under extreme conditions of temperature and pressure generated upon bubble collapse, gas molecules break down into radical molecules which are ions that have unpaired valence electrons. Water molecules in aqueous solutions undergo decomposition to become radical molecules as illustrated in Figure 1.28, below.[58]

Various metal and metal oxide nanoparticles can be prepared *via* sonochemistry. For example, silver nanoparticles can be produced *via* ultrasound irradiation of an aqueous solution consisting of $AgNO_3$ and starch. In this process the starch is cleaved due to the extreme conditions of the reaction

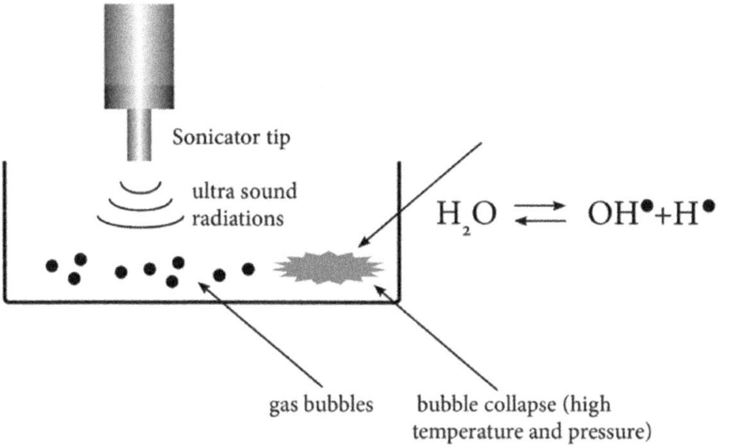

Figure 1.28 Schematic illustration of sonochemical reaction system (reactions under ultrasound irradiation).

and as a result, it reduces the silver ions and prevents them from aggregating. In addition, various ferrofluids have been prepared by sonication including magnetite nanoparticles (Fe_3O_4), nickel, and cobalt nanoparticles.[59–62]

References

1. R. Feynman, *Eng. Sci.*, 1960, **23**(5), 22.
2. M. Ahmad, S. Naseem, M. Arfat, A. Mahmood, K. Qasim and S. Zubair, *J. Biomed. Biotechnol.*, 2014, **11**, 498420.
3. C. B. Michael, *Gold Bull.*, 2004, **37**(1), 12.
4. Y. Yeh, B. Creran and V. M. Rotello, *Nanoscale*, 2012, **4**(6), 1871.
5. R. Marcin, *Chemik*, 2014, **68**(9), 766.
6. P. Stefano, B. Marino, B. Marco, S. Adolfo and F. Giuliano, *J. Mater. Chem.*, 2002, **12**(3), 742.
7. X. Batlle and A. Labarta, *J. Phys. D: Appl. Phys.*, 2002, **35**(6), R15.
8. O. Kostya, C. N. Eric and M. Meyyappan, *Adv. Phys.*, 2013, **62**(2), 113.
9. D. Astruc, F. Lu and J. R. Aranzaes, *Angew. Chem., Int. Ed. Engl.*, 2005, **44**(48), 7852.
10. L. Yongye, L. Yanguang, W. Hailiang, Z. Jigang, W. Jian, R. Tom and D. Hongjie, *Nat. Mater.*, 2011, **10**(10), 780.
11. K. Ibrahim, S. Khalid and K. Idrees, *Arabian J. Chem.*, 2019, **12**, 908.
12. T. Zhiyong, A. K. Nicholas and G. Michael, *Science*, 2002, **297**(5579), 237.
13. C. Harish, *Mater. Sci. Eng., R*, 2005, **49**, 113.
14. S. Steven in *New and Future Developments in Catalysis*: Catalysis by Nanoparticles, Elsevier, Amsterdam, 2013.
15. F. T. Andreas and B. Sotirios, *J. Appl. Phys.*, 2006, **99**(1), 013708.
16. A. P. Alivisatos, *J. Phys. Chem.*, 1996, **100**(31), 13226.
17. V. N. Soloviev, A. Eichhofer, D. Fenske and U. Banin, *J. Am. Chem. Soc.*, 2000, **122**(11), 2673.

18. V. V. Pokropivny and V. V. Skorokhod, *Mater. Sci. Eng., C*, 2007, **27**(5–8), 990.
19. V. Biju, T. Itoh, A. Anas, A. Sujith and M. Ishikawa, *Anal. Bioanal. Chem.*, 2008, **391**(7), 2469.
20. M. S. Dresselhaus, Y. M. Lin, O. Rabin, A. Jorio, A. G. Souza, M. A. Pimenta, R. Saito, G. Samsonidze and G. Dresselhaus, *Mater. Sci. Eng., C*, 2003, **23**(1–2), 129.
21. R. S. Zahra, M. Z. Seyed and K. K. Aniseh, *Nanostructures*, IntechOpen, United Kingdom, 2020, p. 3.
22. K. Mikhail, *Mater. Today*, 2007, **10**(1–2), 20.
23. J. Jaison, B. Ahmed, S. C. Yen, D. Alain and K. D. Michael, *Beilstein J. Nanotechnol.*, 2018, **9**, 1050.
24. D. Johnson-McDaniel, C. A. Barrett, A. Sharafi and T. T. Salguero, *J. Am. Chem. Soc.*, 2013, **135**(5), 1677.
25. W. Philippe, W. Eleonore, H. Philippe, J. Z. Nestor, D. Christopher, C. Jacques, V. Patrick, B. Rene, L. Jean-Luc and T. Georges, *Nano Lett.*, 2006, **6**(10), 2215.
26. B. Heinz, *Chem. Soc. Rev.*, 2007, **36**(1), 15.
27. G. Artioli, I. Angelini and A. Polla, *Phase Transitions*, 2008, **81**(2–3), 233.
28. U. Leonhardt, *Nat. Photonics*, 2007, **1**(4), 207.
29. T. C. Do, P. Annemarie and H. C. Klaus, *Physics and Engineering of New Materials, Springer Proceedings in Physics 127*, Springer-Verlag, Berlin Heidelberg, 2009.
30. S. Delphine and R. Hynd, *Found. Chem.*, 2015, **17**(3), 187.
31. J. H. Florian and N. Markus, *Mater. Today*, 2013, **16**(7–8), 262.
32. H. Horvath, *J. Quant. Spectrosc. RA*, 2009, **110**(11), 787.
33. N. R. Mindy and A. Thomas, *JOM*, 1998, **50**(1), 37.
34. G. Subha and M. Sunit, *Int. J. Nanosci. Nanotechnol.*, 2011, **2**(1), 79.
35. M. G. Payne and M. H. Nayfeh, *Phys. Rev. A*, 1976, **13**(2), 595.
36. G. Binnig and H. Rohrer, *Surf. Sci.*, 1983, **126**(1–3), 236.
37. L. Sumio, *Nature*, 1991, **354**(6348), 56.
38. A. G. Di, *Materials for Ultra-Supercritical and Advanced Ultra-Supercritical Power Plants*, Woodhead Publishing, United Kingdom, 2017.
39. S. A. Zumdhal and S. S. Zumdhal, *Chemistry*, Brooks Cole, USA, 2013.
40. R. F. Egerton, *Physical Principles of Electron Microscopy*, Springer, New York, 2005.
41. S. G. Mokrushin, *Nature*, 1962, **195**(4844), 861.
42. B. Helmut and R. M. Richards, *J. Inorg. Chem.*, 2001, **2001**(10), 2455.
43. G. Dan and E. Matijevic, *New J. Chem.*, 1998, **22**(11), 1203.
44. E. Darrell and D. G. Steven, *General Chemistry*, Brooks Cole, Belmont USA, 2013.
45. J. P. Meyburg and D. Diesing, *J. Chem. Educ.*, 2017, **94**(9), 1225.
46. R. J. Hunter, *Foundation of Colloid Science*, University Press, Oxford, 2001.
47. N. Toshima and T. Yonezawa, *J. Chem.*, 1998, **22**(11), 1179.
48. D. F. Evans and W. Hakan, *The Colloidal Domain: Where Physics, Chemistry, Biology, and Technology Meet*, Wiley-VCH, Canada, 1999.

49. S. Ellerstein and R. Ullman, *J. Polym. Sci.*, 1961, **55**(161), 123.
50. C. Scherer and A. M. Figueiredo, *Braz. J. Phys.*, 2005, **35**(3A), 718.
51. F. Mafune, J. Y. Kohno, Y. Takeda, T. Kondow and H. Sawabe, *J. Phys. Chem. B*, 2000, **104**(39), 9111.
52. L. Rodriguez, M. C. Blanco and M. A. Lopez-Quintela, *J. Phys. Chem. B*, 2000, **104**(41), 9683.
53. P. Grammatikopoulos, S. Steinhauer, J. Vernieres, V. Singh and M. Sowwan, *Adv. Phys.*, 2016, **1**(1), 81.
54. T. J. Mason, *Sonochemistry*, Royal Society of Chemistry, Cambridge, 1990.
55. K. S. Suslick, *Adv. Organomet. Chem.*, 1986, **25**(C), 73.
56. K. S. Suslick, *Science*, 1990, **247**(4949), 1439.
57. M. W. Grinstaff, A. A. Cichowlas, S. B. Choe and K. S. Suslick, *Ultrasonics*, 1992, **30**(3), 168.
58. K. Makino, M. M. Mossoba and P. Riesz, *J. Phys. Chem.*, 1983, **87**(8), 1369.
59. R. Abu-Much and A. Gedanken, *Chem. - Eur. J.*, 2008, **14**(32), 10115.
60. R. A. Mukh-Qasem and A. Gedanken, *J. Colloid Interface Sci.*, 2005, **284**(2), 489.
61. R. Abu-Much and A. Gedanken, *J. Phys. Chem. C*, 2008, **112**(1), 35.
62. R. Abu Much, U. Meridor, A. Frydman and A. Gedanken, *J. Phys. Chem. B*, 2006, **110**(16), 8194.
63. A. Gautam, P. Komal, P. Gautam, A. Sharma, N. Kumar and J. P. Jung, *Metals*, 2021, **11**(2), 329.
64. J. He, N. S. Villa, Z. Luo, S. An, Q. Shen, P. Tao, C. Song, J. Wu, T. Deng and W. Shang, *RSC Adv.*, 2018, **8**(57), 32395.

CHAPTER 2

Teachers' and Students' Awareness of and Attitudes Toward Nanoscience and Nanotechnology

2.1 Introduction

Nanoscience education has increasingly become a central part of the curricula of several universities and high schools around the world. Nanoscience and nanotechnology institutes for science teachers have been established to upgrade high-school teachers' knowledge and understanding of the latest developments in nanoscience.[1,2] In many countries around the world, nanotechnology has been declared a national priority project, and there have been collaborations between academia and local industry and schools. These factors will have a positive impact on the attitudes of teachers and students and their willingness to act positively regarding incorporating the teaching of nanotechnology into the education system.[1–3,9]

The importance of nanoscience and nanotechnology as well as their positive effects on technological and medical developments has required us to increase our awareness and knowledge about it in order to keep up to date with the latest scientific advancements in these fields.[4] Scientists and educators believe that schools are the first starting point.[5] This necessitates creating future generations of global citizens with sufficient scientific literacy and knowledge of leading technological developments in this field. To achieve this goal, as a first step, one should examine the levels of awareness, knowledge and attitudes toward nanoscience and nanotechnology among both teachers and students.[4]

Advances in Chemistry Education Series No. 9
Nanochemistry for Chemistry Educators
By Riam Abu Much, Kurt Winkelmann and Muhamad Hugerat

Published by the Royal Society of Chemistry, www.rsc.org

Teachers' perceptions, beliefs, attitudes and understandings concerning the emerging technologies can influence their approach to science teaching and even their teaching behaviours.[6] In the emerging fields of technology such as nanotechnology, it is necessary to determine teachers' basic knowledge and opinions about these fields so that they can impart the basic required knowledge to their students.[7]

According to Abd-El-Khalick, teachers' perceptions, beliefs, attitudes and knowledge bases concerning emerging technologies can influence their approach to science teaching and even their teaching behaviours.[6] Therefore, in-service teachers as well as pre-service teachers should be informed about different aspects of nanotechnology through appropriate in-service training, seminars, model activities and projects.[1]

In an article by S.-F. Lin *et al.*[8] on science teachers in Taiwan, the impact of science teachers' perceptions and knowledge on nanotechnology and their approach to teaching nanotechnology is clearly shown. Also, the results clearly show that teachers intend to continue professional development and are willing to develop a combination of nanotechnology in teaching and improve knowledge and content in nanotechnology after being exposed to it, especially its connection to daily life.[8]

2.2 Nanotechnology Awareness

Several surveys have shown that the majority of the public have little to no awareness or knowledge of nanotechnology. One study conducted by Waldron, Spencer and Batt among 1500 participants, aged between six and 74 years old, revealed that 60% of them had not heard of nanotechnology at all.[10] They reported that the general public, particularly secondary school students, had a very limited knowledge of nanotechnology. A study carried out by Macoubrie in 2006 was designed to gauge attitudes and opinions of American citizens regarding the development of nanotechnology.[10] Interestingly, this study revealed that 95% of the participants had not heard of nanotechnology before they had learned about it within the context of formal education. A survey carried out by the Australian Office of Nanotechnology indicated that the general public had very high expectations for nanotechnology. However, their concerns were ranked as only moderate, but unfortunately, their knowledge and awareness were rather low.[11,12]

General public surveys about nanotechnology awareness have been carried out in a number of countries, *e.g.*, Japan, USA, UK, Germany, Iran, Italy and Australia.[11] Although the surveys were carried out in different years, ranging from 2004 to 2011, all these countries have high literacy rates. In 2009 Kahan showed in a survey involving 1850 Americans that 81% of the participants had no knowledge about nanotechnology.[13] Similarly, a study conducted in Iran indicated that the majority of the Iranian population are not familiar with nanotechnology. However, those that were familiar with it felt that its benefits outweighed the risks.[14] A study performed by Ho *et al.* suggested that the more that people know about nanotechnology, the more likely they will hold positive attitudes toward it.[15] However, they found that the public perceived nanotechnology as having greater risks and fewer benefits compared to opinions held by nanotechnology experts.[15]

In addition, Scheufele and Lewenstein found that the participants in their study were largely unaware of emerging technologies such as nanotechnology.[16] Furthermore, there was little evidence of knowledge increasing over time. Surprisingly, four in ten Americans indicated that they had heard nothing at all about nanotechnology, and six in ten indicated that they had heard nothing at all about nanotechnology used for human enhancement.

Waldron *et al.* also found that most people have a limited understanding of nanotechnology.[10] Only 60% of their research participants between the ages of 15 and 59 were familiar with the term 'nano' and they associated it with something small. Children under the age of 14 and adults 60 and over were the least familiar with nanotechnology. Their study found that when asked to state the smallest thing they could think of, 52% of 11-year-olds stated microscopic and nanoscopic objects. More elaborate responses were obtained from participants between the ages of 14 and 17 years old. In this group, over 40% chose answers that suggested some knowledge of the nanoscale world, including technical words such as 'an atom,' 'a proton' or 'a molecule'. The greatest level of awareness of nanoscopic objects was evident in young adults from 18 to 22 years of age. Interestingly, adults did not provide more elaborate responses than children about nanotechnology.[10]

A 2014 study conducted by Elmarzugi *et al.* was based on a survey involving many campuses and two governmental research centers (polymer and plastic) in Tripoli, Lybia over a period of about five months. They found that of the 330 participants, 156 knew about nanotechnology but 174 had no knowledge about it.[17] A study conducted by Toqeer *et al.* in Pakistan revealed that the majority of the respondents (77%) had heard about nanotechnology, but only 47% had read about it. A slightly lower percentage (44.4%) had an awareness of nanotechnology applications.[18]

Elki and Sahin investigated Turkish middle-school student awareness, factual knowledge, opinions and risk perceptions toward nanotechnology.[7] Their results indicated that no significant difference was found between students' awareness of nanotechnology based on their gender. A majority of both male and female students consider the potential defense and military applications of nanotechnology to be the most important risk. However, for some of the demographic and affective domain factors, as well as students' achievement in science courses, significant differences were found.[7]

The public's level of awareness with regard to a relatively new technology like nanotechnology will depend critically, among other factors, on (i) the literacy rate of the population, (ii) the awareness created among different segments of the society by the scientific community through different channels of communication, and (iii) the general attitude of the public toward technical innovations. It is therefore important that periodic surveys be carried out to gauge the reaction of the public toward technology in general, and new technologies, in particular. This helps to establish rules and regulations for these technologies as well as to develop policies regarding education, science and technology.[11] The attitude of the public toward nanotechnological products will depend on their awareness of this technology and their perceptions of how these technologies will affect their lives.[19,20]

2.3 Attitudes about Nanotechnology

Regarding the attitudes of students and their teachers with nanoscience and nanotechnology, the influence of mass media was noted. Ho *et al.*'s findings were consistent with those from previous studies in that the mass media shapes public perceptions of nanotechnology.[15] Researchers in the field of communication have argued that the most positive framework of nanotechology in the mass media is likely to act as cues in influencing the public's consideration of both its risks and benefits. In addition, teachers' attitudes toward nanotechnology influenced whether nanotechnology was successfully introduced to K-12 students. Teachers who expressed more positive attitudes toward nanotechnology were more highly motivated to incorporate nanotechnology into their K-12 science programs. Therefore, enhancing positive attitudes in K-12 teachers regarding nanotechnology would be a crucial issue for promoting K-12 nanotechnology education.[15]

2.3.1 Studies about Nanotechnology Attitudes Around the World

In a 2013 study conducted by Balakrishnan, Er and Visvanathan in Malaysia, the responses to their student survey indicated that attaining the objectives of ethical instruction among students was low.[21] They found insufficient awareness among the students about the roles and responsibilities of engineers who can make ethical decisions. The findings of this study also indicated that students' attitudes toward socio-ethical issues related to nanotechnology were low due to less emphasis being placed by the faculty on these issues in an undergraduate engineering program.[21] It was noted from student responses that the nanotechnology engineering program's emphasis was less on sustainable development issues. These issues focused on where faculty members need to incorporate modules that focus on sustainability in technical subjects related to nanotechnology. From the findings of this investigation, it can be concluded that sociotechnical education strongly influences students' knowledge, skills and attitudes regarding socio-ethical issues related to nanotechnology. However, the students in this study were unaware of the engineer's role and responsibilities in society as a whole. This is due to socio-ethical education that is ineffective in building students' knowledge, skills and attitudes regarding socio-ethics (See Chapter 8, Social and Ethical Issues in Nanotechnology).[21,22]

In 2015 Toqeer *et al.*[18] conducted a survey to determine the level of awareness and the attitudes toward nanotechnology among the students and teachers of some higher educational institutions in Islamabad,

Pakistan.[17] The majority of the respondents (77%) had heard about nanotechnology, but only 47% had read about it and a slightly lower percentage (44.4%) had an awareness of its applications. Considering the fact that nanotechnology has not been introduced either as a subject or as a part of the course contents of higher educational institutions that were surveyed, this number represents the percentage of students and teachers who had accessed information about nanotechnology through individual efforts. The level of awareness about nanotechnology increased significantly with the number of years of schooling. They concluded from their survey that a higher level of awareness about nanotechnology and its applications and a higher level of education had a positive impact on the attitude of the participants toward nanotechnology.[18]

An additional study conducted by Ekli and Sahin investigated the awareness, factual knowledge, opinions and risk perceptions of students from Turkish middle schools concerning nanotechnology in a very general sense.[4] It was carried out on 1396 middle-school sixth, seventh and eighth-grade students. Student perceptions and opinions about nanotechnology were elicited through a questionnaire developed by the authors. It was found that students had some awareness of nanotechnology, and that most of them had positive feelings and opinions about it. Students' perceptions of nanotechnology and the influence of their demographic and affective domain factors, achievement in science courses. and science motivation were also investigated. The results indicated that for gender no significant difference was observed. However, significant differences were found for some of the demographic and affective domain factors, along with achievement in science courses.[4]

2.4 Attitudes about Nanotechnology – The Israeli Case

Abu-Much *et al.*'s study examined the awareness and attitudes of teachers and students from the Arab communities in Israel toward nanotechnology.[23] Both teachers' and students' knowledge and attitudes were examined as a function of different factors, including gender, grade, level of education and the number of years of experience. The participants in this study were randomly selected teachers and students from the Arab sector in Israel. The participants were asked to complete a questionnaire regarding their knowledge and attitudes toward nanotechnology. The study was adapted from the Norwegian Relevance of Science Education Project (ROSE) questionnaire except that the concepts of science and technology were replaced by nanotechnology and nanoscience.[24] The participants were asked to answer 14 questions concerning different aspects related to nanotechnology. The majority of the students and teachers knew that nanotechnology deals with small things. A high number also indicated its importance in medical and technological applications. The lowest number was found for teachers and students who had no idea about nanotechnology.[23]

The Abu-Much study[23] indicates that the majority of teachers and students were aware of the importance of nanotechnology for medical and technological applications, whereas a lower percentage did not know of its importance. Students generally have shallow knowledge about nanotechnology and its applications. However, this shallow knowledge does not reflect their true knowledge of it. The students had heard of the word nanotechnology but they did not have any technical understanding of it. It was expected that these students would not have had any definitive opinion about nanotechnology. This could reflect the inferior quality of the teaching or the lack of information about nanotechnology in the curriculum.[23]

Abu-Much's results[23] indicate that teachers' attitudes were slightly more positive toward nanotechnology. Their attitudes were more positive than students' attitudes and their attitudes were considered to be moderately positive. The results also indicate that there were no differences in teachers' attitudes toward nanotechnology that were related to gender, school, educational level or years of experience. In addition to the quantitative data assembled *via* questionnaires, a semi-structured questionnaire interview was conducted for five randomly selected teachers and five students. According to the participants' answers in this interview, both the teachers and students lack considerable information about the field of nanotechnology.

Abu-Much *et al.*'s study indicates that in the Israeli Arab population, both teachers and students have only a basic or have insufficient knowledge about nanotechnological concepts. Interestingly, nobody specifically referred to the nanometre scale. In investigating the attitudes of teachers and students toward nanotechnology, Abu-Much *et al.* found that their responses were at a low level, which indicated that they had very superficial or insufficient knowledge.[23]

Other findings in the literature were found to be consistent with the results of the research conducted by Abu-Much and his team. Toqeer *et al.* found that most respondents had heard about nanotechnology but only 47% had read about it.[18] A slightly lower percentage (44.4%) of respondents had an awareness of its applications. According to research done by Hingant and Albe, students from Turkish middle schools had some awareness of nanotechnology and most students expressed positive opinions about it.[25] A study conducted by Sheetz and her team indicated that only 17% of the respondents knew what nanotechnology was. This research[26] found that males had a higher percentage *vs.* females, which is inconsistent with Abu-Much *et al.*'s results.

A research study conducted in Germany that examined public attitudes toward and awareness of nanotechnology also revealed that most of the participants were not familiar with nanotechnology.[27] Research was carried out in Singapore regarding public perceptions of nanotechnology.[28] About 80% of the respondents had "some" understanding of nanotechnology, 60% reported having heard some negative information about it and 39% perceived nanotechnology as beneficial. Ekli and Sahin found that science and technology teachers in Turkey had 'moderate' knowledge of nanotechnology and that they learned about it mostly from their daily lives.[7] Their knowledge was found to be inadequate, whereas the survey participants in the Ekli and Sahin study had positive attitudes toward nanotechnology.[7] Abu-Much *et al.*'s results[23] also agree with those of Elmarzugi *et al.*,[17] who found that many Libyan students displayed a favourable attitude toward nanotechnology.[16] This was reflected in the teachings about nanotechnology in the Libyan undergraduate and postgraduate curriculum.

2.5 Conclusion

In many countries around the world, nanotechnology has been declared a national priority project, and there have been collaborations between academia and local industry and schools. These factors will have a positive impact on the attitudes of teachers and students and their willingness to act positively regarding incorporating the teaching of nanotechnology into the education system.

The importance of nanoscience and nanotechnology as well as their positive effects on technological and medical developments has required us to increase our awareness and knowledge about it in order to keep up to date with the latest scientific advancements in these fields. Teachers who expressed more positive attitudes toward nanotechnology were more highly motivated to incorporate nanotechnology into their science programs.

References

1. F. Allhoff, P. Lin, and and D. Moore, *What Is Nanotechnology and Why Does It Matter? From Science to Ethics*, John Wiley & Sons Ltd, United Kingdom, 2009.
2. B. Bhushan, *Springer Handbook of Nanotechnology*, Springer-Verlag, Berlin Heidelberg, 2010.
3. M. G. Wacker, *J. Pharm. Sci.*, 2014, **103**(3), 777.
4. E. Ekli and N. Sahin, *Int. J. Technol. Des. Educ.*, 2013, **23**(4), 867.
5. R. T. Putnam and H. Borko, *Educ. Res.*, 2000, **29**, 4.
6. F. Abd-El-Khalick, *J. Sci. Teach. Educ.*, 2001, **12**(3), 215.
7. E. Ekli and N. Sahin, *Procedia Soc. Behav. Sci.*, 2010, **2**(2), 2667.
8. S.-F. Lin, J.-Y. Chen, K.-Y. Shih, K.-H. Wang and H.-P. Chang, Science teachers' perceptions of nanotechnology teaching and professional development: a survey study in Taiwan, *Nanotechnol. Rev.*, 2015, **4**(1), 71–80.
9. S. Joachim, *Scientometrics*, 2004, **59**(3), 425.
10. A. Waldron, D. Spencer and C. Batt, *J. Nanopart. Res.*, 2006, **8**(5), 569.
11. J. Macoubrie, Nanotechnology: public concerns, reasoning and trust in government, *Public Underst. Sci.*, 2006, **15**(2), 221–241.
12. AON, Australian Office of Nanotechnology, Department of Industry, Tourism and Resources (DITR), Canberra, www.industry.gov.au, (accessed 25 September 2007).
13. D. M. Kahan, *Nat. Nanotechnol.*, 2009, **4**, 705.

14. P. Farshchi, S. Sadrnezhaad, N. Nejad, M. Mahmoodi and L. Abadi, *J. Nanopart. Res.*, 2011, **13**(8), 3511.
15. S. Ho, D. Scheufele and E. Corley, *Sci. Commun.*, 2011, **33**(2), 167.
16. D. Scheufele and B. Lewenstein, *J. Nanopart. Res.*, 2005, 7, 659.
17. N. A. Elmarzugi, E. I. Keleb, A. T. Mohamed, H. M. Benyones, N. M. Bendala, A. I. Mehemed and A. M. Eid, *J. Appl. Pharm. Sci.*, 2014, **4**(1), 110.
18. A. Toqeer, I. Saba, Y. Khawaja and M. R. Sayed, *J. Nano Educ.*, 2015, **8**(1), 44.
19. M. C. Roco, *Nat. Biotechnol.*, 2003, **21**, 1247.
20. D. M. Berube, *Nano-Hype: The Truth Behind the Nanotechnology Buzz*, Prometheus, USA, 2005.
21. B. Balakrishnan, P. H. Er and P. Visvanathan, *Sci. Eng. Ethics*, 2013, **19**, 1341.
22. R. McGinn, Ethics and Nanotechnology: Views of Nanotechnology Researchers, *Nanoethics*, 2008, **2**, 101–131.
23. R. Abu-Much, A. Basheer, S. Zahaika and M. Hugerat, *Create Educ.*, 2019, **10**, 1140.
24. The ROSE project: An overview and key findings, http://roseproject.no/network/countries/norway/eng/nor Sjoberg-Schreiner-overview-2010.pdf.
25. B. Hingant and V. Albe, *Stud. Sci. Educ.*, 2010, **46**(2), 121.
26. T. Sheetz, J. Vidal, T. D. Pearson and K. Lozano, *Technol. Soc.*, 2005, **27**(3), 329.
27. R. Asmatulu and H. Misak, *J. Nano Educ.*, 2011, **3**(1–2), 13.
28. M. C. Roco, *J. Nanopart. Res.*, 2011, **13**, 427.

CHAPTER 3

Sharing Nanotechnology with Students

3.1 Introduction

In recent years huge strides have been made in nanotechnological advances. Basic scientific innovations and practical developments have brought new challenges to academia (See Chapter 6). As a result, many science education systems around the world have revised their curricula to offer relevant nanoscience and nanotechnology education courses. All of these converging phenomena highlight the need to properly educate engineering and science students and equip them with the ability to design, analyse and synthesise nanosystems.[1]

Some of the important questions in the field of contemporary science teaching are as follows:

- Why do we choose to introduce the subject of nanoscience in high school?
- Is it just to satisfy the next generation of nano scientists?

Science teacher educators understand that perceptions regarding the nanoscale held by teachers will consequently affect the perceptions of their students (See Chapter 2). It can be assumed that professional development programmes could have implications as to what teachers consider most important to teach in their classrooms. This may affect how teachers will teach and how students will learn nanoscience concepts.

Nowadays, implementing nanotechnology education for schools is a common practice around the world. Technology education can use nanoscience and nanotechnologies to stimulate dialogue about important issues, for example, regarding the relationship between science, technology

Advances in Chemistry Education Series No. 9
Nanochemistry for Chemistry Educators
By Riam Abu Much, Kurt Winkelmann and Muhamad Hugerat

Published by the Royal Society of Chemistry, www.rsc.org

and society. Since nanotechnology is now a reality, and not science fiction, it should also be included in higher education curricula. It is imperative to find appropriate and natural ways to incorporate nanotechnology into school science, technology, engineering and math curricula. However, integrating essential concepts of nanoscience, traditional science and technology into middle-school science courses may be particularly challenging.[1]

Research from Sakhnini and Blonder was designed to identify entry points of the eight Nanoscience and Nanotechnology (NST) concepts essential to science and technology encompassing nanoscience ((1) size-dependent properties, (2) innovations and applications of nanotechnology, (3) size and scale, (4) characterization methods, (5) functionality, (6) classification of nanomaterials, (7) fabrication approaches of nanomaterials, and (8) the making of nanotechnology) in order to introduce these subjects into middle-school science and technology curricula in Israel.[2]

In this research,[2] middle-school science and technology teachers took a course that included all eight essential concepts for nanoscience, science and technology in order to help them understand the essential concepts in these subjects in depth. Then, they were asked to identify the point of entry in the existing scientific and technological curriculum for each of the essential concepts of nanotechnology, science and technology. To further validate this research, two different groups of teachers participated in two consecutive phases of the study: the identification phase and the validation phase. The teachers in the identification phase identified the entry points of all eight essential concepts of science, nanotechnology and technology in the science and technology curriculum. This reflects the relevance of the concepts of science, nanotechnology and technology from the teacher's point of view in terms of pedagogical approach. Most of the identified integration points were verified in the second stage. Forty-two intersection points of concepts essential to science, nanoscience and technology were proposed to be integrated into the science and technology curricula. All of the points offered at the identification stage were approved by a second set of middle-school teachers at the verification stage.[2]

At the verification stage, 11 new points were added. The categorisation of these points are described as follows: 19 income points were offered to be integrated during the teaching of chemicals in chemistry curricula; 12 in the life sciences, four for the physics and energy-related subjects and seven in technological systems and products. The results of this study showed that multiple exposure opportunities could be offered to middle-school students through their existing science and technology curriculum. The study also serves as an example of how other schools could integrate concepts of science, nanoscience and technology into the existing science curriculum in Israeli middle schools and around the world (See Chapter 4).[2]

In their research, Ban and Kocijancic present examples of innovative methodologies for teaching nanotechnology education including ongoing activities, experiments, computer-supported programs, materials for students and teachers and so on.[3] The investigators proposed including some topics related to nanotechnology in middle-school curricula within the existing engineering and compulsory technology subjects and in one of the elective subjects. The proposed experiments were considered simple to perform and required only resources that are generally widely available.[3]

Nanoscience and nanotechnology can be presented to students studying science and technology *via* teachers preparing student activities such as experiments and demonstrations (see examples in Chapters 5 and 6). This can be achieved by encouraging a 'science day' that would focus on nanoscience and nanotechnology. The ideal solution would be to create a sustainable initiative to introduce engineering and technology education into the school. The initiative could be dedicated to modern nano-related technologies with a focus on relevance to everyday life. An important part of the 'science day' initiative would be to hold courses for teachers who deal with and teach nanotechnology and prepare them for introducing the positive advancements that nanotechnologies provide.[3,4] In several countries around the world, various institutions and museums offer exhibitions (a 'nano day') to schools and the general public. The same organisations arrange student group visits to educational institutions that research and develop nanoscience and nanotechnology topics. There are also workshops, seminars, interactive lectures and many online resources that provide information on nanoscience and nanotechnology topics. Many organizations around the world prepare events and activities to celebrate the field of nanotechnology through the "nano day", including how it is currently used in products that enrich our daily lives, and the challenges and opportunities it holds for the future in our daily life.

There has been substantial development of nanoscience-based research and the widespread use of nanotechnological methods in both academic research and industry worldwide. One of the main goals of developing curricula around the world is to develop and educate future generations of nanoscience researchers. This can occur when there is more of an emphasis on the history of nanoscience education, including curricula that will be integrated into educational institutions for all ages. The field of nanoscience and nanotechnology education has immensely gained in importance in recent years because of the widespread use of nanotechnology in everyday life.[5] To encourage and promote nanoscience education, it is very important that a short history of the growth of nanoscience education is presented in the curricula timeline. Introducing the elements of history of science seems to contribute to students' understanding of concepts and methods and to the creation of positive attitudes towards science.[5]

It is also very important to implement nanoscience education projects within the schools in order to properly expose students to nanoscience. Learning through projects proceeds from the premise that there is a topic or question that needs to be researched and studied, a process that comes to fruition and presentation in the product, which should be meaningful to the person and the world. During the project, students acquire a deep understanding of academic knowledge and mastery of learning skills, acquiring skills such as teamwork, brainstorming, information retrieval and presentation to an audience. They prepare high-end products that are presented to an audience that includes experts.[5]

The first article in the *Journal of Chemical Education* that included nanotechnology as a topic appeared in 1995.[6] The subject quickly penetrated the scientific education system in the United States and around the world. In 2004, a national centre was established in the United States called The National Centre for Learning and Teaching in Nanoscale Science and Engineering (NCLT). This centre consists of self-described, integrated, nano-educational component areas including learning research, nano concepts, courses, learning technology development, professional development, resource dissemination, networking and community building, as well as evaluation and assessment. The centre prepares curricula for all ages on nanochemistry and nanotechnology concepts and it makes these topics relevant to students' social life. The project is named 'Big Ideas' and it is designed to teach nanoscience to grades 7–12 and grades 12–18. The project emphasises effective teacher professional development programmes for including nanoscience in school curricula.[7]

3.2 Teaching High-school Students Nanoscience and Nanotechnology

It is important that today's students in current educational systems become more familiar with nanotechnology. It is therefore imperative to present the subject in the context of chemistry education in school and to adequately train science teachers. To support this, specific study materials and teaching methodologies on nanotechnology have been developed.[8–10]

Many researchers have identified and noted nine 'Big Ideas' (Size and Scale, Structure of Matter, Forces and Interactions, Quantum Effects, Size-Dependent Properties, Self-Assembly, Tools and Instrumentation, Models and Simulations, and Science, Technology and Society) related to nanoscience and nanotechnology. Based on these Big Ideas, a learning sequence can be developed for middle-school students; it focuses on size and scale, tools and instrumentation, size-dependent characteristics and a scientific society.[11,12]

In a study by Stavrou *et al.*,[11] the Big Ideas teaching sequence was used in a class of 15 high-school students in eighth grade, ages 14–15 years old. Seven meetings were held, and each lasted about ninety minutes. The results indicated positive encouragement in the teaching of science, nanotechnology and technology.[11] The findings from Stavrou *et al.*[11] made it possible to determine that the Big Ideas process can improve students' understanding of nanotechnology and their awareness of social and ethical issues related to nanotechnology, science and technology (see Chapter 4).[11–13]

Blonder and Dinur introduced a nanotechnology study module designed for high-school students.[14] The module presented a unique way of adapting advanced content to the level of high-school students through constructivist pedagogy where students were at the centre of learning. The study elaborated on the idea of the transition from a teacher-focused pedagogy to a student-focused pedagogy. The student-cantered pedagogy provided a positive, ongoing motivation for the students. In addition, another study by Blonder indicated that combining applied modules of nanotechnology using appropriate instructional methods in high-school science classes could provide unique solutions for enhancing low enrolment percentages of students who attend class in science subjects.[8]

Ter Horst *et al.*[10] designed a nanotechnology teaching module built according to a guided learning concept. It was designed for chemistry classes in high schools with reference to standard German educational standards at that time.[18] However, it could be adapted to other countries educational standards. Their nanotechnology module focused on zinc oxide particulate matter particles in four parts (How Does "Nano" Work; How to Produce "Nano"; Why Do You Need "Nano"; "Nano: Risks *versus* Potentials").[10]

Many of the teachers confirmed that they would use this teaching module at their school even though the topic seemed quite demanding. Teachers also rated the unit as refreshing because it offered new and interesting topics to teach chemistry education and these topics were also related to modern research.[10] This indicates that there was interest in this topic that had not yet been addressed during regular science classes. Teachers stressed the relevance of the topic and that it contributed to meeting many educational standards in the regular curriculum. Therefore, it seems worthwhile to introduce the subject to schools and to develop new teaching concepts related to nanoscience and technology.[10,15]

Velentzas and Stavrou developed and implemented a teaching and learning sequence (TLS) for high-school students that focused on carbon nanostructures.[16] Because of the great importance of the subject of nanotubes, and especially carbon pipes, this topic was chosen. Although carbon pipes and nanotubes have already been used in a wide variety of applications, there are impressive indications of their use for additional, innovative, future applications. It is very important to keep citizens informed about important emerging scientific issues in our technological world.

Velentzas and Stavrou[16] prepared a module in which students study and examine the various geometric structures of a C_{60} fullerene 'Bucky ball', also known as single-walled carbon tubes. This module was prepared in order to introduce nanotechnology to the students *via* presenting some of the interesting features at the nano level, which are related to the structure of the material. While watching a video, students could learn about what nanotechnology is, how scientists in the field of nanoscience work, and some products and applications of nanotechnology that are used in everyday life. Students could watch a video in parts and a teacher-guided discussion could accompany each part.[16]

3.3 AFM to Nano

Thanks to the development of tools such as the atomic force microscope (AFM), nowadays scientists and technologists are able to not only observe but also manipulate nanoscale sized objects. Since the new technological revolution, we have witnessed the existence of two different technological approaches: top-down and bottom-up. They greatly contribute to raising useful discussions concerning the definition of nanoscience and nanotechnologies.[17,18] Developments in nanoscience from top to bottom refer to miniaturisation. The opposite technological approach, from the bottom up, refers to building objects by assembling molecules or aggregates. This approach refers to the opposite direction of miniaturisation: it starts with molecules that form larger objects.

Many years ago, the atomic force microscope (AFM) was developed by quantifying the short-term interaction of van der Waals interactions. By using an AFM, one can study biological samples down to atomic resolution. Information on topography, roughness, friction, adhesion, elastic properties, the interaction between edge and sample surface, the distributions of electric field, magnetic field, resistance, surface potential and more can all be obtained by the AFM at nanometre resolution.[18]

Planinšič and Kovač[18] presented an innovative model of teaching while using an AFM. This model was proven successful as an introduction to nanophysics for high-school students as well as for physics teachers. The model demonstrated the two operating modes of the AFM (touch mode and oscillation mode) as well as some basic principles that limit the resolution of the method. This model could be used in the classroom during class as a demonstration experiment, a simple lab experiment or students could have access to an AFM at home.[19] This could easily serve as a good model for introducing the subject of nanoscience and nanotechnology into the school curriculum.

3.4 STEM and Nanotechnology Topics

Various arguments justify the introduction of nanoscience and nanotechnologies in classrooms. Innovative articles and ideas have dealt with teaching or accepting nanotechnologies as a socio-technological issue. Zenner and Crone proposed activities for teaching nanotechnology and societal issues at the middle-school level. Since then, many programs around the world have included training graduate students to bring nanotechnology to the public.[20]

Many scientists emphasise the importance of introducing nanoscience topics into educational curricula when preparing intervention programmes in STEM on the subject of nanoeducation for all student types and ages. This is in preparation for encouraging and educating future nanoscientists and helping to lessen the global lack of nanoscientists and nanotechnologists.[7,21]

In preparing intervention programmes in STEM on nanoeducation, one must take three points into account: consideration of the essential interdisciplinary nature of nanoscience itself; dealing with the latest standards and scientific concepts in the field of nanotechnology; and the curricula should include the topic of nanoscience in society. These three points are of particular importance in designing relevant curricula and developing resources or pedagogical assessments in nanoscience. Schank *et al.*[22] argued that this educational emphasis on nanoscience will serve as a catalyst that will lead to a substantial modernisation of the scientific educational system for high-school students worldwide. Nanoscience and nanotechnologies are required for updating standards, for creating connections between traditional disciplines, and ultimately for advancing true interdisciplinary teaching that conforms to the modern realities of science education.[22,23]

Therefore, many education systems around the world are working hard to ensure that nanoscience-educated workers are adequately and properly trained. Adding more workers skilled in nanotechnologies has become urgent for two reasons. First, most industrialised countries are fiercely competing among one another to lead in the number of scientific and engineering advancements. In addition, the number of students currently choosing STEM careers, their academic skill sets, and their scientific literacy levels, especially in nanoscience, are considered insufficient.

3.5 Nanoscience, Nanotechnology and Society

Ethical and social aspects of nanoscience and nanotechnology must be taken into account at the university level.[24,25] In addition, the difficulties encountered, and decisions made with regard to content and people's needs should be considered in nanoscience course development and teaching. In 2006 Sweeney was able to identify and analyse how researchers grasp the social and ethical dimensions of a 'nano work' summer programme (see Chapter 8).[25]

The University of Wisconsin at Madison in the USA developed a course entitled 'Nanotechnology and Society'. It was designed and led by graduate students themselves or it was taught to them in another course called 'Nanoscience and Nanotechnology in Society'. The course was based on encouraging open discussion among students from various nanotechnology fields.[26] It is very important that social and ethical implications should be integrated into university science courses to ensure that future scientists do not neglect them (see Chapter 8).[27,28]

Researchers Toumey and Baird preferred the term "interactions" over implications to introduce the idea of the joint development of nanotechnology and society.[28] They wanted various initiatives taken at the University of South Carolina to contribute to "Building nanoliteracy in the university". They described various programmes and activities from undergraduate courses to outreach programmes in order to discuss social and ethical interactions of nanotech with laymen.

3.6 Conclusion

It is very important to note that many scientists, educators and members of society claim that nanotechnologies open up new possibilities for developing electronics, new materials, medicine, chemistry, pharmaceuticals, biotechnology, agriculture and more. Therefore, it is very important to discover and share examples of innovative materials and scientific methods in nanoscience with students of all ages and make nanoscience a part of the current school curricula.

References

1. S. Ozel and Y. Ozel, presented in part at Proceedings of the 5th WSEAS/IASME International Conference on Engineering Education, Heraklion, Greece, July 2008.
2. S. Sakhnini and R. Blonder, *Nanotechnol. Rev.*, 2018, 7(5), 373.
3. K. Ban and S. Kocijancic, presented in part at The Second World Conference on Technology and Engineering Education, Ljubljana, Slovenia, September 2011.
4. A. Laherto, *Sci. Educ. Int.*, 2010, **21**(3), 160.
5. A. E. Greenberg, *ACS Nano*, 2009, **3**, 762.
6. L. A. Coury Jr, M. Johnson and T. J. Murphy, *J. Chem. Educ.*, 1995, 72, 1088.
7. S. Y. Stevens, L. Sutherland, P. Schank and J. Krajcik, *The Big Ideas of Nanoscale Science and Engineering*, NSTA Press, Arlington, Virginia, 2009.
8. R. Blonder, *J. Nano. Educ.*, 2010, **2**, 67, 9.
9. J. H. Tomasik, S. Jin, R. J. Hamers and J. W. Moore, *J. Nano. Educ.*, 2009, **1**(1), 48–67.
10. N. Ter Horst, T. Wilke and T. Waitz, presented in part at New Perspectives in Science Education, Conference Proceedings, Florence, Italy, 2015, March, 2015.
11. D. Stavrou, E. Michailidi, G. Sgouros and K. Dimitriadi, Teaching high-school students nanoscience and nanotechnology, *LUMAT Int. J. Math Sci. Technol. Educ.*, 2015, **3**(4), 501.
12. S. Swarat, G. Light, E. J. Park and D. Drane, *J. Res. Sci. Teach.*, 2011, **48**, 512.
13. A. Magana, S. Brophy and L. Bryan, *Int. J. Sci. Educ.*, 2012, **34**(14), 2181.
14. R. Blonder and M. Dinur, *J. Nano. Educ.*, 2012, **3**, 1.
15. A. Velentzas and D. Stavrou, *Chem. Teach. Int.*, 2021, 3(1), 45.
16. A. Velentzas and D. Stavrou, *Chem. Teach. Int.*, 2021, 3(1), 45.
17. D. Vinck and Les Nanotechnologies, ed. *Le Cavalier Bleu*, Paris, France, 2009.
18. G. Planinšič and J. Kovač, *Phys. Educ.*, 2008, **43**, 37.
19. K. E. Drexler, *Phys. Educ.*, 2005, **40**, 339.
20. G. M. Zenner and W. C. Crone, in *Nanoscale Science and Engineering Education*, ed. A. E. Sweeney and S. Seal, American Scientific Publishers, Valencia, California, USA, 2008, pp. 621–647.
21. L. A. Bryan, S. Daly, K. Hutchinson, D. Sederberg, F. Benaissa and N. Giordano, presented in part at The Annual Meeting of the National Association for Research in Science Teaching, New Orleans, USA, April, 2007.

22. P. Schank, J. Krajcik and M. Yunker, in *Nanoethics: The Ethical and Social Implications of Nanotechnology Can Nanoscience Be a Catalyst for Education Reform?* ed. F. Allhoff, P. Lin, J. Moor and J. Weckert, Wiley Publishing, Hobeken, New Jersey, USA, 2007, pp. 277–289.
23. P. Hurd, *J. Res. Sci. Teach.*, 2002, **39**, 3.
24. R. W. Berne, *Nanoscale Science and Engineering Education*, ed. A. E. Sweeney and S. Seal, American Scientific Publishers, Valencia, California, 2008, pp. 547–566.
25. A. Sweeney, *Sci. Eng. Ethics*, 2006, **12**, 435.
26. C. Tahan, R. Leung, G. Zenner, K. Ellison, W. Crone and C. A. Miller, *Am. J. Phys.*, 2006, **74**, 443.
27. C. A. Miller and S. K. Pfatteicher, in *Nanoscale Science and Engineering Education*, ed. A. E. Sweeney and S. Seal, American Scientific Publishers, Valencia, California, 2008, pp. 567–576.
28. C. Toumey and D. Baird, in *Nanoscale Science and Engineering Education*, ed. A. E. Sweeney and S. Seal, American Scientific Publishers, Valencia, California, USA, 2008, pp. 577–589.

CHAPTER 4

Incorporating Nanochemistry in the Chemistry Curriculum

4.1 Introduction

How teachers introduce students to a topic matters as much as what they teach. Educators have developed many strategies, each with their benefits and deficiencies. Like every other subject, there is no single course structure for teaching nanotechnology that works for all situations. At the same time, the educational benefits of student-centeredness, active participation, and authentic learning experiences used for teaching other science disciplines apply to nanotechnology as well. Before implementing programs described in this chapter, readers should read the primary literature so that they replicate the pedagogy as well as the activities themselves.

This chapter begins with a description of the challenges that middle and high school faculty face when trying to add nanotechnology to their curriculum. Whenever possible, suggestions are provided that can minimize the time, effort, and expense that teachers and schools face with such a project. It then presents implementation frameworks that demonstrate success based on research studies. In some cases, untested methods for improving student learning are included as suggestions.

4.2 Challenges of Adding Nanotechnology to the Chemistry Curriculum

Teachers willing to include cutting edge science topics like nanotechnology face numerous potential impediments. It does not matter how many interesting and educational activities are available if teachers have no way to present them to students. While these barriers to improving the chemistry

Advances in Chemistry Education Series No. 9
Nanochemistry for Chemistry Educators
By Riam Abu Much, Kurt Winkelmann and Muhamad Hugerat

Published by the Royal Society of Chemistry, www.rsc.org

curriculum are real and common, there are resources and strategies available that can minimize the difficulties.

4.2.1 Curriculum Insertion Points

Most importantly, the chemistry curriculum is already full or even overloaded (more details are presented in Chapter 6). This is apparently true around the world, based on the many sources of such comments. As society becomes increasingly dependent on technology, science, and engineering, there is little hope that science standards will relax so that they contain fewer learning outcomes and curriculum requirements. Teachers might be even more discouraged when they observe that their country's or region's standards for science education might not even mention nanotechnology.

Educators have addressed this concern by finding nanotechnology topics that are consistent with current science standards. These topics are referred to as insertion points. By selecting a series of specific insertion points, a teacher can feature nanotechnology throughout the year's curriculum. Alternatively, a broad topic such as how science influences society, can serve an insertion point for a larger module about nanotechnology (*e.g.*, ethics, applications). Science standards vary among and within countries as well as over time, so teachers will need to adjust the insertion points described here to suit their specific situation.

The nine Big Ideas of nanotechnology are one way to organize the discipline's content.[1] They are:

- Size and scale,
- Structure of matter,
- Forces and interactions,
- Quantum effects,
- Size-dependent properties,
- Self-assembly,
- Tools and instrumentation,
- Models and simulations, and
- Science, technology, and society.

There is no requirement to introduce all nine Big Ideas. Some Big Ideas are more closely related to engineering, physics, biology, or sociology, although all touch upon the central science of chemistry. When choosing, instructors should realize that some Big Ideas build upon others. For instance, the first four Big Ideas are interrelated, spontaneous self-assembly is due to attractive forces and interactions among nanostructures, and measurements using tools and instrumentation often depend on quantum effects. Thus, by learning about a series of related Big Ideas, students can build upon their prior nanotechnology knowledge and see its interconnectedness. The following paragraphs provide the reader with a brief explanation of the importance of each of the Big Ideas. More detailed information is provided in Chapter 1, Nanotechnology and Chemistry: The Unseen Scale with Magnificent Impact.

Size and scale for nanotechnology encompasses dimensions, shape, geometry, and related quantities such as surface area, volume, and the surface area to volume ratio. All these measurements are critically important to understanding the features of nanomaterials (see the lab activity described in Section 6.4, Chapter 6). Measurement scales of distance, mass, and time are fundamentally important for all science disciplines. Working with measured quantities requires mathematical reasoning and conceptual understanding of size regimes. It is likely that chemistry students have already learned the prerequisite knowledge for size and scale understanding in mathematics and prior science classes. Some instruction might be necessary for understanding scales at the atomic level (*i.e.*, nanometres) but this would be necessary regardless of whether the chemistry instructor includes nanotechnology in the curriculum.

The structure of matter describes the existence of atoms, ionic units and molecules, and how those particles interact with each other. As with size and scale, this is foundational knowledge for science, especially chemistry. Nanotechnology provides some unusual examples, such as allotropes of carbon. Diamond, graphite (including individual layers called graphene), fullerenes, and carbon nanotubes all consist entirely of carbon but their different structure leads to different properties (more detailed information is provided in Section 1.3 in Chapter 1).

Interactions among the building blocks of matter further determine the properties at the atomic, nano-, and macroscale. These include chemical bonding as well as intermolecular forces. The inherent instability of nanoparticles themselves can be related to the lack of opportunities for bonding among atomic units on the surface of the nanoparticle. Interactions at the nanoscale, such as those between nanoparticles with surface coatings of ions, are explained by the Coulombic attractive and repulsive forces.

The mathematical complexity of quantum mechanics certainly prevents a full quantitative understanding at the middle and high school (or even graduate school) level. However, a qualitative description of the wave–particle duality, the Heisenberg Uncertainty Principle and quantized energy levels can be sufficient. The next step is to extend these ideas to applications. The colour change from bulk to nanosized particles and tunnelling for scanning tunnelling microscopy are obvious examples of quantum effects observed in nanotechnology. Quantum effects in nanotechnology are visually appealing, providing an easy way to excite students' curiosity.

Besides colour changes, many properties are size-dependent at the nanoscale. Chemical and physical properties in general are important topics within chemistry but examples are almost exclusively reserved for bulk materials with only a mention that some properties are very different for individual atoms or molecules. In this case, nanotechnology could help students make more sense of this difference. Instead of treating properties of atomic units as being separate and distinct from properties of bulk materials, the size-dependent properties show that the property's value lies on a spectrum related to the particle's dimensions.

Students should be familiar with self-assembly from their experiences in biology, although they might not immediately make the connection. The helical structure of DNA and arrangement of phospholipid molecules in a cell membrane are both examples of important self-assembly processes. Consequences of forces and interactions between molecules are often presented in chemistry, such as the impact of hydrogen bonding on water. Nanoscale dimensions are still small enough that Coulombic attractions can dominate, making nanostructures larger scale examples of how intermolecular forces affect the behaviour of particles. By extending a discussion about intermolecular forces, an instructor can include monolayer films and micelles. A thorough presentation of the importance of self-assembly in a biological setting is found in Chapter 7, Nanoliposomes as a Model for Teaching Nanochemistry.

Curiously, almost all of chemistry as we know it was built upon just indirect evidence of atoms. It was not until the late 20th century that scientists "saw" atoms. Even then, the image of atoms on a surface was generated from measurements of an electrostatic repulsive force or quantum tunnelling current between a probe tip and a surface through atomic force microscopy (AFM) or scanning tunnelling microscopy (STM), respectively. Publicly available STM and AFM images of atoms and molecules can be interesting ways to present atoms to students when they learn about the particulate nature of matter. They can learn about AFM when discussing electrostatic forces between atoms. The repulsion between the surface and probe tip and the repulsion that keeps two atoms separated to form a bond are the same phenomenon. Chapter 5, Activities for Teaching Nanotechnology, provides specific examples of teaching nanotechnology using AFM.

The interpretation of the physical universe, a central purpose of science, requires the use of models. Predicting behaviour of any system requires a simulation that is built upon a model. Therefore, students should already have some experience using models and simulations in prior science classes and for other topics in chemistry. Like atoms and molecules, nanoscale structures are too small to see directly, so models are an essential component of many presentations of nanochemistry.

Incorporating the last Big Idea, the interdependence of science, technology and society, is becoming increasingly common at all levels of education. Students are more familiar with and often more interested in technological applications of science and their implications on their community, rather than science itself. Near-future applications of nanotechnology and the societal impacts, such as the positive and negative effects on the environment and ethical uses of nanotechnology, are discussion topics that can increase students' interest in the scientific aspects. If it is possible to coordinate instruction in different subjects, then chemistry students could engage in activities described in Chapter 8 during their civics or government class sessions and connect that learning experience with the scientific lessons taught in the chemistry class.

The Big Ideas are not the only way to categorize nanotechnology concepts. Other education researchers have developed an alternative list that includes: size-dependent properties, innovations and applications, size and scale, characterization methods, functionality, nanomaterial classification, and the making of nanotechnology.[2] Using these frameworks, teachers have found insertion points for nanotechnology topics throughout the middle school and high school chemistry curriculum.[1,3,4] For instance, when introducing the size and scale of atoms, teachers can highlight that the nanoscale lies between the atomic size (0.1 nm) and the size range of objects seen through a typical microscope (1 μm). Teachers can illustrate the usefulness of most chemistry topics with an example of a product or process that involves an application of nanotechnology. In fact, there are many opportunities to incorporate nanotechnology into the middle and high school curriculum. Readers are encouraged to review the insertion points described in detail within ref. 1, 3, and 4.

In surveys about the challenges of teaching nanotechnology, teachers point out that they teach specific disciplines, such as chemistry, but nanotechnology is an interdisciplinary subject. While this is true, most of the chemical phenomena demonstrated by nanomaterials reside completely within the range of topics covered in a typical chemistry class. Some knowledge of environmental science, medicine, electronics, and engineering is necessary to appreciate the diverse applications of nanotechnology, but that is true for the many ways that chemistry is used for technological innovations as well. Teachers can incorporate as much information about other disciplines as necessary to spark students' interest, but chemistry can still be at the heart of the nanotechnology concept.

Discussions of how teachers relate an academic subject like nanotechnology to the interests of students often relies on presenting the science "in context". By providing context, students can connect the information that they will learn with ideas that they believe are meaningful. The addition of context adds to the time spent preparing and teaching a lesson but simply stating that a scientific concept has some useful application does not guarantee that students will learn more. It is important that teachers understand how to do this in a way that improves student learning.[5] The approaches to teaching subjects in context here and in Chapter 6, Nanochemistry as a Relevant Concept in Teaching Chemistry, are consistent although the latter is a more thorough treatment of this important issue.

Context in relation to chemical or nanotechnology education can be presented through the temporal (past or future), location, and social descriptions of a focal event. Students should feel that the focal event is significant to such an extent that they wish to be involved in a community of practice that is based on that event. Reasons of significance include a social concern, job training, or a connection to an aspect of their everyday lives. For instance, studying photocatalytic TiO_2 nanoparticles as a disinfectant for water can be useful as a focal event to put chemistry in context. Water treatment is familiar to students and they may understand the significance of clean drinking water in developing countries. By studying this application, they participate in a community interested in improving the quality of life for people worldwide. A teacher should present the focal point at the beginning of the lesson and as much as possible reintroduce the focal point (or add new details) throughout the lesson.

The context should relate directly to the topic being discussed. A teacher could use the context of TiO_2 nanoparticles to teach laboratory skills learned through the nanoparticle synthesis, size and scale, size-dependent properties, and redox reactions. As with any activity, the teacher should reasonably expect students to achieve the learning objectives. This requires that the teacher understand the students' background knowledge of both the context and the scientific topics. Related to this is the problem that by weaving in a non-chemistry topic, students will encounter familiar words that have a different meeting. For instance, the word "light" is necessary when discussing photocatalytic TiO_2 nanoparticles. In common usage, light means the visible spectrum but TiO_2 is white, reflecting all visible light. How can light be involved with TiO_2? Instead, TiO_2 is useful because it absorbs light (meaning electromagnetic radiation) in the UV region of the spectrum. Instructors must clearly differentiate word meanings that are found in the context and the scientific concept.

The lesson should allow students to contribute their own ideas about the focal event. To continue the example, students could be asked whether to use TiO_2 nanoparticles in water disinfection, given that the nanoparticles might be released into the environment. Successfully incorporating context into a lesson will enable students to learn the scientific concepts through talking about the focal event. Even better, introducing a series of related focal events will allow students to relate ideas from one focal point to another. For instance, students could use their knowledge about TiO_2 nanoparticles for water disinfection to discuss silver nanoparticles used in medical bandages. Descriptions of activities that relate to these nanotechnology applications can be found in Chapter 5, Activities for Teaching Nanochemistry.

4.2.2 Equipment, Curriculum, and Professional Development Resources

Chemistry educators may lack equipment and curriculum materials needed to introduce nanotechnology into their classes. These are long-standing issues that limit the proliferation of nanotechnology education. However, educators are making some headway by creating and sharing resources.

Equipment limitations may not be as severe as a new nanotechnology teacher might fear. Almost all activities presented in Chapters 5 and 8 use only common chemistry lab supplies. Chemicals for nanoparticle synthesis can include gold and silver salts but there are many other excellent experiments using less expensive chemicals. The only uncommon pieces of equipment are STM and AFM instruments and even education-grade versions are expensive. Fortunately, alternatives are available. Students can use models to simulate microscopy data collection (see Chapter 5, Section 5.3.1). Teachers can partner with a university that offers remote access to instrumentation. Depending on the instrument capabilities, students watch a researcher collect the data or they perform the experiment using online controls. They may be able to choose the sample to be analysed. In either case, students can receive the data to view themselves offline. A third option is to share the cost to purchase a piece of equipment with other schools and rotate access among many teachers.

Chemistry educators consistently cite availability of curriculum resources as an impediment to teaching nanotechnology. Resources do exist, as this this text demonstrates. However, teachers need to know where to look. The *Journal of Chemical Education* frequently publishes nanotechnology-themed experiments. The *Journal of Nano Education*, published from 2009 to 2016, includes many experiments appropriate for a chemistry course. Online databases of activities are maintained by the National Nanotechnology initiative, the NANOYOU project, and NanoHUB.[6–8]

Arguably, time is a teacher's most limited resource. Teaching nanotechnology is just now becoming common in the undergraduate science curriculum. Many teachers have no exposure to nanotechnology, making it difficult for them to even know where to begin. Teachers frequently participate in professional development events to keep themselves abreast of new education policies, pedagogies, and instructional tools. Opportunities exist for teachers to learn about nanotechnology as well.

For teachers who wish to learn on their own, high-quality educational resources are available at websites such as NanoHUB,[8] and many books are available to introduce readers to nanotechnology. The primary literature cited in Chapter 5 provides background theory about the chemical phenomena presented by the activities.

4.3 Learning Nanotechnology Through Writing, Multimedia, and Art

Students reinforce their knowledge by communicating it to others. Students can plan their presentation, perform research, organize information, practice writing or public speaking, and possibly use slideshow and video editing software. Written, oral, or artistic presentations allow students to express themselves and demonstrate knowledge in an authentic manner. Group projects help students develop cooperation and teamwork. Gaining these real-world skills benefit students in future academic and work environments.

The most common way to incorporate communication skills into the curriculum is for students to research a topic and give an oral presentation to their class. Presenting nanotechnology concepts to a class of students in a lower grade can benefit both the older presenting students and the younger audience. In one example, high school students completed a lesson about nanotechnology used in chemical sensors that included a laboratory activity and viewing experts' presentations online. Since the topic was too new for it to appear in textbooks, students found information through their research. Working in groups, students prepared a presentation about the lesson to be given to middle school students. This required the presenters to not only understand the information well enough to present it. They had to adjust their presentation by considering the background knowledge of the less academically advanced audience. This approach increased nanotechnology content knowledge of the high school students and they retained that knowledge even after several intervening months.[9]

A collaborative project involving college nanotechnology and art students could be adapted to a high school setting. After learning about scanning electron microscopy (SEM) and its ability to achieve nanometre resolution of surfaces, science students in an introductory nanotechnology course were paired with students enrolled in a three-dimensional design art course (more detailed information about electron microscopy is provided in Chapter 1).[10] Each pair of students acquired SEM images and the art students made a clay relief of one image then prepared a plaster cast. The science students prepared a book of the SEM images. Both the book and plaster casts were presented and remained on display in the art department for several days. This is a creative way for students to learn about nanotechnology – science students helped art students learn about SEM, students could explore the artistic aspects of nanoscale images, and students created a tangible product that was viewed by their peers. It is not necessary for high school students to acquire their own SEM images since many are available online. STM and AFM images could provide artistic inspiration as well.

College faculty have developed other innovative ways for students to express their artistic creativity and understanding of nanotechnology. These activities could also fit within a high school or middle school chemistry curriculum. In addition to lectures, video clips showing nanotechnology in popular culture, and laboratory experiments, students studying game design, cinema art, fiction writing, and other liberal arts subjects prepare a final project about a nanotechnology topic of their choice. The mode of presentation has included video videos, live singing, short stories, movies, graphic art, and a comic book.[11]

Comic books are an example of a medium that appeals to teenagers and can contain information in an educational and entertaining way. Educators have published *Knowing Nanotechnology via Comics*, a comic book about nanotechnology.[12] They compared student learning of nanotechnology after assigning the comic book or a booklet written in standard scientific textbook prose.[13] Both media covered applications, size and scale, size-dependent properties, instrumentation, forces and interactions, and societal impacts of nanotechnology. The booklet was ten pages and the comic book was ten times longer. Low, medium, and high academic achieving students preferred learning *via* the comic book rather than the booklet. Overall, all three groups improved their understanding of nanotechnology regardless of the whether they read the booklet or comic, with learning gains proportional to the students' overall academic ability. However, the highest achieving group of students learned more through the booklet than the comic. The opposite was true for the medium ability students. Low performing students learned equally well using either. These results suggest that although students might prefer learning in a certain way, their choice may not be in their best interest.

Multimedia is also a useful way for experts and instructors to convey information to students. With video teleconferencing becoming more common, teachers can more easily arrange presentations by nanotechnology experts. In one case, high school chemistry students learned about metal–organic frameworks (MOFs), which contain nanoscale pores that make them useful in medicine and water purification.[14] To study the impact of the presentation delivery method, some students viewed a YouTube video or a slideshow, both prepared by MOF researchers. The video was recorded and edited using inexpensive consumer hardware and software. In some cases, one of the MOF researchers was present in the classroom. Researchers asked students to complete surveys about their learning about chemistry applications, the job of a chemist, authentic laboratory work, and miscellaneous topics. Results of students' self-reported learning gains shows that most thought that they learned something from both types of presentations, even when the research chemist was not present to answer questions. Students learned more about authentic lab research and the job of a chemist from the video.

University students demonstrate their understanding of nanotechnology through communicating with outside audiences just like high school students do. Several studies have shown that non-scientific audiences, such as high school students, can learn about the nanotechnology in this way.[15,16] The audience responds favourably when presenting students craft their delivery in an engaging, interesting manner.[16] Blogs provide another way for experts to disseminate nanotechnology information in an informal manner to a general audience.[17] The writers gain experience communicating their knowledge with non-scientists. Since blogging and other social media platforms are free and easy to use, high school students could use them to learn about nanotechnology or use them to demonstrate their own learning gains.

4.4 Learning Nanotechnology in Extracurricular Settings

Service learning is becoming popular as a means for students to gain authentic learning experiences. A student engages in a community service-related activity outside of a classroom, connects that experience with academic knowledge and skills, communicates the outcomes to a community audience, and reflects upon how the overall experience relates to their career goals and interests. Students gain a realistic perspective of how to apply their academic knowledge in the real world. By reflecting on the service learning, students can revise and refine their college and career goals. Examples of college students engaging in service learning as part of a chemistry or engineering course are found in the literature.[18–20] Many of these ideas can be adapted to secondary school classes. For instance, high school students can present nanotechnology demonstrations at science outreach events or develop nanotechnology activities that elementary and middle school students perform.

Science-themed summer day-camps and workshops immerse middle and high school students in an informal, fun learning environment. Camps held at universities can include lab tours, meetings with science experts, and presentations about preparing for college.[21] All of these activities can inspire students to study STEM subjects in college. Science camps are fundamentally different than classrooms during the school year because participation is voluntary and camp leaders have a great deal of freedom when structuring the activities. Since participants are giving up a portion of their summer, they are more likely to have a genuine interest in science. This chapter author and another Chemistry faculty member have led Camp Nano, a nanotechnology summer day-camp for high school students. Camp Nano participants performed many of the experiments described in Chapters 8 and 9, including measuring electrical current with a simulated STM, building a Lego® AFM, imaging materials with education-grade scanning probe microscopes, synthesising ferrofluids, preparing Au nanoparticles in plastic, and debating the ethics of nanotechnology.

Students from a low-performing public middle school have attended a summer science camp in which they performed a twelve-hour unit about size and scale.[22] The context of this project was about antimicrobial properties of nanomaterials. Students worked in groups as an advisory board investigating the use of nanotechnology to eliminate germs from surfaces. They viewed hair and bacteria cells under a microscope and learned about relative sizes of different objects. For their final activity, students created a video presentation about the usefulness of nanotechnology to prevent bacteria outbreaks at school. After completing these activities, camp participants' understanding of certain size and scale information exceeded that of high school students and matched the knowledge of students attending a private middle school. This demonstrates the value of a well-designed extra-curricular activity.

A two-day summer workshop designed by GE and Columbia University teaches high school students about materials science and nanotechnology, with an emphasis on learning about industrial processes, product scale-up, and real-world trade-offs between technical and economic considerations of a product.[23] In the first half of the workshop, called a training block, students learn skills, knowledge, and attitudes that enable them to complete the project-based learning educating block.

The training block includes lectures, discussions, and hands-on activities. It begins with lessons about the multidisciplinary nature of materials science and the societal impacts of nanotechnology. The next lesson focuses on material properties that are needed to achieve product performance goals. They are introduced to nanotechnology-enabled products, such as thin films containing gold nanoparticles that change colour under certain conditions. In the third session, students learn about examples of nanostructures found in nature, designing materials based on designs inspired by nature (biomimicry), and methods used for measuring material properties.

Students work in groups during the educating block to study the properties of different types of candy. Although this activity does not involve nanomaterials, students use their knowledge of material science and engineering gained during the training block to develop their own testing methods and perform measurements. In response to a post-workshop questionnaire, students expressed a stronger commitment to pursuing a career in science or engineering. They were more confident in their knowledge of materials properties and appreciated the trade-offs between product performance and cost.

4.5 Instructional Strategies for Integrating Nanotechnology into the Chemistry Curriculum

There are two general approaches for adding a new subject into an existing curriculum. The first is to continuously mention it throughout the course, perhaps by replacing old activities with new nanotechnology course material. The NanoTeach program[24] and the Bumpy/Shaky/Sticky framework[25] are examples of continuous inclusion of nanotechnology in a chemistry course. The second method is to make space in the class schedule for an intensive, multi-lesson unit about nanotechnology, much like a covering a chapter in the textbook. Scale Worlds[26] and NanoLeap[27] programs are good examples of using a modular approach to teach nanotechnology. The NanoSense program was designed to accommodate either integration method.[28] This topic is briefly discussed here and is treated in a more in-depth manner elsewhere.[29]

Regardless of the desired approach, it is important to not further overload the curriculum. Teachers should review their course material and determine if there are any lessons that can be shortened, combined, or eliminated while still meeting all necessary science standards. Decide which nanotechnology topics students should learn. If a crowded curriculum remains a concern, select topics that fit easiest within the existing curriculum so less time is needed to bridge the gap between students' existing knowledge and the new nanotechnology topic. Section 4.2.1 discusses how nanotechnology concepts overlap with chemistry. Any new activities or lessons should have a clear connection to larger concepts of chemistry and align with the required education standards.

Revising the chemistry curriculum gives teachers an opportunity to try new pedagogies and different types of activities. Educators are developing innovative teaching methods for nanotechnology, including game-based learning, storytelling, and others.[30] If a teacher removes an expository chemistry lab activity, an inquiry-style nanotechnology experiment updates the activity and improves student learning. Any new lesson will be more educational if it presents the nanotechnology topic in a proper context that interests students and connects to something meaningful in their lives. Faculty seeking to implement new nanotechnology activities should first gain the appropriate background knowledge themselves through professional development opportunities. They should evaluate their students' learning, followed by refining the new course material as needed.

4.6 Conclusions

As educators develop new chemistry-related nanotechnology activities, they have also studied how to best incorporate them in or outside of class. New instructional and communications technology make this task easier for teachers. A variety of effective methods exists that teachers can choose based on their needs. Barriers to incorporating new topics remain but solutions exist to minimize the time and effort required. Revising and updating a curriculum gives teachers an opportunity to improve their teaching methods, by introducing better pedagogy and more up-to-date information.

References

1. S. Y. Stevens, L. M. Sutherland and J. S. Krajcik, *The Big Ideas of Nanoscale Science and Engineering*, NSTA press, Arlington, Virginia, 2009.
2. R. Blonder and S. Sakhnini, *Global Perspectives of Nanoscience and Engineering Education*, Springer International Publishing, AG Switzerland, 2016, pp.117–127.
3. S. Sakhnini and R. Blonder, *Nanotechnol. Rev.*, 2018, **7**, 373.
4. R. Blonder and S. Sakhnini, *Chem. Educ. Res. Pract.*, 2017, **18**, 903.
5. J. K. Gilbert, *Int. J. Sci. Educ.*, 2006, **28**, 957.
6. Educational Resources for K-12 Teachers, https://www.nano.gov/education-training/teacher-resources, (accessed May 2021).
7. Nanoyou, https://nanoyou.eu/index.html, (accessed May 2021).
8. NanoHUB Education, https://nanohub.org/groups/education, (accessed May 2021).
9. P. Ambrogi, M. Caselli, M. Montalti and M. Venturi, *Chem. Educ. Res. Prac.*, 2008, **9**, 5.
10. A. Bentley and G. Imatani, *J. Nano Educ.*, 2013, **4**, 33.
11. K. Kostecka, *J. Nano Educ.*, 2016, **8**, 133.
12. National Changhua Normal University, *Knowing Nanotechnology via Comics, Ehgbooks*, Changhua, Taiwan, 2017.
13. S.-F. Lin and H.-S. Lin, *Int. J. Sci. Educ.*, 2016, **38**, 1373.
14. S. R. Burgin, Y. Sakamaki, M. Tsuji, O. Watson, Z. Heidrick, T. Chitwood, M. Benamara, E. M. Martin, M. Childress and M. H. Beyzavi, *J. Chem. Educ.*, 2020, **97**, 2351.
15. J. Strachan, C. Barnett, T. Maschmeyer, A. F. Masters, A. Motion and A. K. L. Yuen, *J. Chem. Educ.*, 2020, **97**, 4166.
16. S. K. St Angelo, *J. Chem. Educ.*, 2018, **95**, 804.

17. L. M. Bishop, A. S. Tillman, F. M. Geiger, C. L. Haynes, R. D. Klaper, C. J. Murphy, G. Orr, J. A. Pedersen, L. DeStafano and R. J. Hamers, *J. Chem. Educ.*, 2014, **91**, 1600.
18. K. J. Donaghy and K. J. Saxton, *J. Chem. Educ.*, 2012, **89**, 1378.
19. J. Duffy, L. Barrington, W. Moeller, C. Barry, D. Kazmer, C. West and V. Crespo, *Int. J. Service Learn. Eng*, 2008, **3**, 18.
20. A.-M. L. Nickel and J. K. Farrell, Using Service Learning to Teach Students the Importance of Societal Implications of Nanotechnology, in *Service Learning and Environmental Chemistry: Relevant Connections*, ed. E. S. Roberts-Kirchhoff, M. J. Mio and M. A. Benvenuto, ACS Symposium Series, American Chemical Society, Washington, DC, 2014.
21. J. Donnelly, C. Diaz and F. E. Hernandez, *J. Chem. Educ.*, 2016, **93**, 619.
22. C. Delgado, S. Y. Stevens, N. Shin and J. Krajcik, *Nanotechnol. Rev.*, 2015, **4**, 51.
23. L. Avila and L. Fine, *J. Nano Educ.*, 2010, **2**, 27.
24. D. Huffman, J. Ristvey, C. Morrow and M. Deal, in *21st Century Nanoscience - A Handbook*, ed. K. D. Sattler, Public Policy, Education, and Global Trends, CRC Press, Boca Raton, 2021, **vol. 10**, pp. 10.5–10.7.
25. T. R. Tretter, in *21st Century Nanoscience - A Handbook*, ed. K. D. Sattler, Public Policy, Education, and Global Trends, CRC Press, Boca Raton, 2021, **vol. 10**, pp. 11.5–11.7.
26. T. R. Tretter, *Nanotechnol. Rev.*, 2015, **4**, 33.
27. E. A. Palmer, *The NanoLeap Project Evaluation Report 2007-2008*, ASPEN Associates, Inc., 2009.
28. P. Schank, A. Wise, T. Stanford and A. Rosenquist, *Can High School Students Learn Nanoscience? An Evaluation of the Viability and Impact of the NanoSense Curriculum*, SRI International, 2009.
29. D. Huffman, J. Ristvey, C. Morrow and M. Deal, Integrating Nanoscience in High School Science: Curriculum models and Instructional Approaches, in *21st Century Nanoscience - A Handbook*, ed. K. D. Sattler, Public Policy, Education, and Global Trends, CRC Press, Boca Raton, 2021, **vol. 10**, pp. 10.1–10.9.
30. R. Blonder and S. Sakhnini, *Chem. Educ. Res. Pract.*, 2012, **13**, 500.

CHAPTER 5

Activities for Teaching Nanochemistry

5.1 Introduction

A typical chemistry curriculum describes the behaviour of individual atoms and molecules, bulk properties of materials, and the connections between the two. As described elsewhere in this book, nanotechnology lies in between the size scales of the atom and the macroscopic world. Although the development of nanotechnology is relatively recent, its impact on science, engineering, and our daily lives is already significant. The economic forecast for nanotechnology is strong, so it is imperative that teachers introduce students to this topic within their science courses.

While nanotechnology is not entirely within the domain of chemical sciences, many topics taught in a traditional chemistry class – atomic structure, bonding, reactions, and material properties – are fundamental to understanding nanotechnology as well. Societal issues discussed within a chemistry class, such as recycling and clean energy, can be addressed through the applications of nanotechnology research. Therefore, a high school chemistry class or a unit about chemistry in a middle school physical science class is a natural home for nanotechnology. Biology, physics, and integrated science classes can also include nanotechnology topics due to the interdisciplinary nature of nanotechnology. This chapter focuses on the aspects of nanotechnology best suited for incorporating into a chemistry curriculum.

Fortunately, there exists a wide variety of nanotechnology learning activities for middle and high school chemistry students. This chapter presents just a representative sample of those resources and in no way is it intended to be a comprehensive or exhaustive review of activities. Activities described here were designed for high school students and found in the peer-reviewed chemical education literature. Activities have been tested in

Advances in Chemistry Education Series No. 9
Nanochemistry for Chemistry Educators
By Riam Abu Much, Kurt Winkelmann and Muhamad Hugerat

Published by the Royal Society of Chemistry, www.rsc.org

the high school classroom unless noted otherwise. Teachers of advanced high school chemistry classes might consider looking for experiments designed for introductory college courses. An effort is made to select activities from different countries to highlight the diversity of educational approaches used around the world. Methods for incorporating nanotechnology activities into courses is presented in another chapter so that topic is not discussed in any depth here. Also, it is assumed that the teacher will decide how much introductory information to provide to students and the manner of presenting it. This chapter focuses on describing what students will do and what they will learn when the activity is underway.

This chapter's selection of activities encompasses many of the Big Ideas of Nanoscale Science and Engineering,[1] (henceforth referred to as Big Ideas) which are:

- Size and scale,
- Structure of matter,
- Forces and interactions,
- Quantum effects,
- Size-dependent properties,
- Self-assembly,
- Tools and instrumentation,
- Models and simulations, and
- Science, technology, and society.

Background information about the Big Ideas and other frameworks for teaching nanotechnology are also discussed in Chapter 4, Incorporating Nanochemistry in the Chemistry Curriculum. A single activity may encompass multiple Big Ideas. The following section of this chapter presents activities that allow students to synthesize nanomaterials and demonstrate their applications. Section 5.3 presents activities with hands-on manipulation of models, including those that explain scanning probe microscopy. In many cases, simulations of complex topics allow instructors to highlight the desired concepts without distracting students with extraneous information or procedures. The use of models eliminates the need for expensive instrumentation or materials and allows younger students to learn about nanotechnology safely. Activities related to the impact of nanotechnology on society are presented in Chapter 8.

Hands-on activities, especially wet chemistry experiments, are most often published in the chemical education literature. Among those, the lower-cost experiments specifically identified for high school and middle school are highlighted here. Teachers can modify most experiments to be a demonstration rather than a hands-on activity if necessary. The original procedures of many activities described below are expository rather than inquiry- or discovery-based, so teachers may wish to modify them accordingly. The use of research instrumentation and the synthesis of certain nanomaterials may not be appropriate for students, but teachers can present these topics through remote observations and demonstrations. Activities presented here can substitute for a more traditional chemistry lab activity. See Chapter 4, Incorporating Nanochemistry in the Chemistry Curriculum, for information about incorporating these topics.

5.2 Nanomaterial Synthesis and Applications

Before presenting a selection of interesting and educational nanochemistry experiments, a few recommendations/warnings are necessary. Since this chapter does not fully describe experimental procedures, interested readers must consult the cited articles and their supporting information before attempting them. If an article is not freely available, curious teachers should contact the activity's author directly to ask for a copy. Abide by all appropriate safety and environmental regulations and policies. The original sources often contain safety guidelines, learning objectives, assessment tools, student worksheets, instructors' notes, and other useful information.

Unless otherwise noted, all experiments should be performed by groups of two or three students rather than individually. This approach reduces the amount of chemicals, supplies, and equipment used for the activities. Some activities are sufficiently difficult that a team of students is necessary in order to complete their work. Constructivism, the most prevalent learning theory in chemical education, acknowledges the importance of the social context in which students learn,[2] especially in the laboratory.[3] The practice of students working in pairs or small groups in lab promotes persistence, academic performance, and attitudes related to science subjects.[4] Collaboration encourages creative thinking[5] which is required for successful completion of inquiry activities. All of these reasons for students working in small teams apply to the activities described in this chapter.

Within this section, experiments are organized by nanoparticle type. This allows instructors to create activities for one or multiple lab sessions with each experiment building upon another by using the same concepts and materials. All experiments in this section demonstrate the Big Ideas of structure of matter, forces and interactions, and size-dependent properties. Whenever possible, the activities emphasize the applications of these nanomaterials, so that students can learn about the impact of nanotechnology on technology and society.

5.2.1 Titanium Oxide and ZnO Nanoparticles in Sunscreen

Sunscreen is a cosmetics product familiar to middle and high school students that can contain nanoparticles. Due to their small size, TiO_2 and ZnO nanoparticles can absorb harmful UVA and UVB light while still being transparent. Larger particles would scatter longer wavelengths of light, creating a white paint-like coating on the wearer's skin. The nanoparticles are mixed into a cream and sold to consumers. Two experiments highlight the use of nanomaterials in sunscreen. In the first, students prepare a nanoparticle-infused,

UV-blocking cream and in the second experiment, they can isolate nanoparticles from a commercially available sunscreen product.

5.2.1.1 Preparation and Evaluation of Nanosized ZnO-based Sunscreen

The first experiment is noteworthy because it builds up from nanoparticle formation through their use in a common consumer product, touching upon several important chemical topics along the way.[6] Students can observe the size-dependent properties of ZnO nanoparticles during the preparation of their own sample of sunscreen. Nanoparticles form in an emulsion, which is a common synthesis method for nanoparticles that relies on intermolecular forces (hydrophobic/hydrophilic interactions) among solutes and solvents. They measure the impact that the different sized particles' surface area to volume ratios have on their ability to absorb UV light. Students prepare a sunscreen-like material in an emulsion and again evaluate its UV light protection. Although other methods are available for students to measure the ability of sunscreen to block UV light,[7–9] this method is the simplest and requires no equipment other than a black light. Besides nanotechnology, students can learn about emulsions and fluorescence in this activity. Teachers could incorporate this experiment into a lesson about chemistry improving our health (*i.e.*, preventing sunburns and possible skin cancer).

Students begin by preparing nano- and microscale $Zn(OH)_2$ particles in order to compare the influence of size on their behaviour. Aqueous solutions of zinc chloride and sodium hydroxide are mixed and quickly form a cloudy solution of macroscopic zinc hydroxide particles which precipitate. To form nanoparticles, students perform the same reaction but in a solution of heptane with sodium dioctyl sulfosuccinate (commonly referred to as AOT). Although the synthesis requires a mixture of organic components, the instructor can prepare this part of the solution ahead of time to minimize their use by students.

A surfactant possesses a charged functional group at one end of the molecule that interacts favourably with an aqueous phase and a hydrocarbon "tail" that is stable in an organic phase. In the presence of both aqueous and organic liquids, surfactant molecules undergo self-assembly by spontaneously organizing themselves into nanosized, loosely formed, spherical structures. The phase present in the lesser amount is held within the structure, separated from the majority phase. If the phase within the sphere is organic, these structures are called micelles. The charged functional group faces outward towards water with the tail pointed inward. Soap forms micelles in water to separate the oil (dirt) phase from the surrounding water solvent. When the organic phase is in the majority, as is the case for this experiment, then the structures are referred to as reverse micelles with the surfactant tail pointing outward.

Students prepare reverse micelles containing droplets of aqueous $ZnCl_2$ and NaOH solutions within heptane. This is also called a water-in-oil (w/o) emulsion (more information is provided in Appendix 1, Chapter 7). Upon stirring, the small AOT reverse micelles collide, causing the AOT molecules to yield and allow mixing of the droplets. Unlike in the simple aqueous solvent where all the Zn^{2+} and OH^{-} can interact with each other, the size of the droplets is limited by the reverse micellular structure of the AOT molecules around them. The reaction of zinc chloride and sodium hydroxide occurs in each droplet, producing $Zn(OH)_2$ particles ranging in size from 1 to 100 nm. Due to their small size and the stability of the w/o emulsion, the $Zn(OH)_2$ nanoparticles do not agglomerate and precipitate. This is the students' first clue that although they mixed the same reactants to form the same product, the restricted growth of the $Zn(OH)_2$ particles in the emulsion leads to a fundamentally different result.

Students isolate both types of $Zn(OH)_2$ particles from their respective solvents. Heating at 60 °C dehydrates $Zn(OH)_2$ to form ZnO but this step is not necessary. Both $Zn(OH)_2$ and ZnO will behave similarly in all subsequent steps of the experiment.

Another size-dependent property of the ZnO particles is the extent that they absorb UV light. The absorption of UV light increases with the total surface area since only surface units can absorb light. Although a smaller particle has less surface area than a larger particle (assuming both are spherical), a collection of smaller particles will have greater surface area compared to an equal volume (or mass) of larger particles. This means that the ratio of surface area to volume increases for a given mass of particles as the size of individual particles decreases. The surface area to volume ratio is an important property in nanotechnology. In this case, given equal amounts of smaller or larger particles, the smaller ZnO particles absorb more UV light.

Students can easily observe this effect. They first colour two spots with a yellow highlighter on white paper. The highlighter ink fluoresces when illuminated by UV light from a black light. They dissolve equal masses of micro- and nanosized ZnO powders in glycerine then spread the liquids on separate highlighted spots. Shining the black light on the highlighted spot covered with ZnO nanoparticles results in a dark spot appearing instead of bright fluorescence from the ink. In contrast, the fluorescence is only partially obscured if the highlighter ink is covered with larger ZnO particles.

Following this, students can prepare a material similar to sunscreen by mixing ingredients that form a basic skin cream. They add equal masses of zinc oxide nanoparticles or microparticles to separate samples of skin cream and repeat the highlighter spot test. The cream containing particles with higher surface area to volume ratio will absorb more of the UV light. Although this is a similar procedure to the previous steps, it is significant for students to observe the benefit of nanoparticles in a material that is like the sunscreen product sold commercially.

5.2.1.2 Isolation of TiO_2 Nanoparticles from Sunscreen and Observation of Their Chemical Properties

Titanium dioxide is a versatile compound used in many applications. Its white colour makes TiO_2 useful as a brightener for cosmetics and paper products. Like ZnO, TiO_2 nanoparticles in sunscreen are capable of absorbing UV light. In fact, it is possible to separate the TiO_2 nanoparticles from commercially available sunscreen then use them in other chemical reactions. That is the goal of the following multipart experiment.[10]

The yield of titanium dioxide depends on the UV protection provided by the sunscreen, so a higher SPF value is desirable. Isolation of TiO_2 from sunscreen begins with overnight heating of the cream at 120 °C to remove water and low-boiling organic compounds. The remaining material is calcined to drive off any remaining organic components. This second step could be performed by a teacher prior to class or as a demonstration. The original publication of this experiment describes several methods for confirming that the residual white powder is in fact titanium dioxide nanoparticles. While instrumental analyses are impractical for students to perform themselves, teachers may view this as an opportunity to discuss characterization techniques such as scanning electron microscopy and X-ray diffraction. Raw data from such analyses are provided in the supporting information of ref. 10. The authors found that the TiO_2 nanoparticles had an average diameter of 40 nm, although this may differ depending on the sunscreen studied.

Although it is not discussed in the original version of the experiment or tested, it is possible that this TiO_2 powder could be used in the same way as ZnO nanoparticles to demonstrate their ability to absorb UV light. Students can disperse the TiO_2 nanoparticles in glycerine and in the simple skin cream described previously. This could be a convincing way for students to observe the presence of UV-absorbing TiO_2 nanoparticles.

Titanium dioxide can remove organic pollutants from water using UV light.[11] It is a UV photocatalyst – a material that catalyses reactions when it absorbs UV light. Upon the absorption of a photon, a TiO_2 particle can remove pollutant molecules on its surface through a combination of oxidation and reduction reactions. Since the TiO_2 is losing electrons through oxidation of a pollutant and gaining electrons through a concurrent reduction reaction, the TiO_2 returns to its original state at the end of the reaction, ready to catalyse the degradation of another organic molecule.

Students can perform a simple photocatalytic reaction using TiO_2 nanoparticles isolated from sunscreen. The model pollutant is nontoxic methylene blue dye in water. The decomposition of methylene blue requires both UV light and TiO_2. Therefore, students should prepare two control or reference solutions containing (1) just methylene blue and (2) methylene blue and TiO_2. This second solution should be covered to prevent exposure to UV light. A third aqueous solution of TiO_2 and methylene blue and the control solution 1 are exposed to UV light. Within minutes, students will observe that the blue colour of solution 3 (methylene blue $+TiO_2+$ UV light) has faded while exposure to UV light has not changed the blue colour of solution 1 (methylene blue only). The presence of TiO_2 nanoparticles in the non-irradiated solution 2 is also insufficient to cause a colour change.

Another use of the sunscreen-derived TiO_2 nanoparticles is as an anti-microbial agent. Titanium dioxide is acting as a photocatalyst but in this case it is reacting with biological cells rather than individual organic molecules. Students add yogurt (a microorganism source) and milk to four beakers. Titanium dioxide is added to solutions 3 and 4. Solutions 2 and 4 are exposed to UV light. Solutions 1, 2, and 3 are controls because they lack (1) TiO_2 nanoparticles and UV light, (2) TiO_2 nanoparticles, and (3) UV light. Only solution 4 experiences all the conditions necessary to prevent the growth of mould. All four beakers are covered to prevent foreign bacteria from growing. Several days later, mould is visible in beakers 1 and 2 which lack TiO_2. Mould growth occurs in beaker 3 but it is partially inhibited by TiO_2 absorption of stray UV light (*e.g.*, sunlight). Beaker 4 remains free of any microorganisms.

Teaching applications of TiO_2 nanoparticles in context is discussed in Chapter 4, Incorporating Nanochemistry in the Chemistry Curriculum.

5.2.2 Copper(I) Oxide Particles

Nanoparticles exhibit size-dependent optical properties due to the change in the number and arrangement of their energy levels. Figure 5.1 illustrates this change. A single atom or ionic unit absorbs a discrete amount of energy as electrons move from the lower-energy ground state to a higher-energy excited state (a). Interactions between individual atoms or units within a cluster cause slight changes in the energies of the ground and excited states. The minimum energy difference between the ground and excited states decreases as the cluster grows from a few units (b) to thousands (c). At this

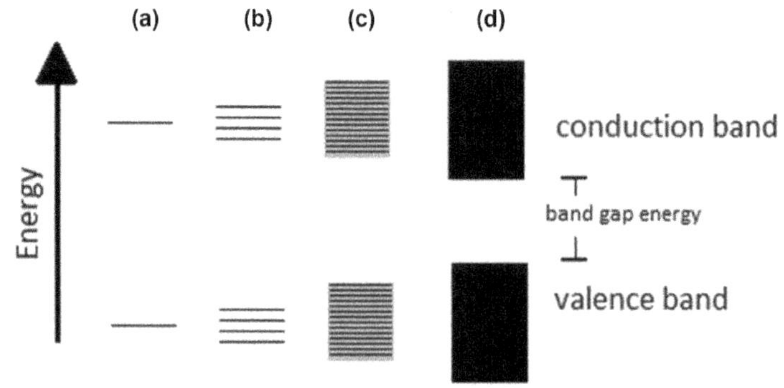

Figure 5.1 Diagram showing changes in the energies of excited and ground states as an atom (a) grows into a small cluster (b), a nanoparticle (c), and finally into the bulk material (d).

point, a nanoparticle has formed. Its continued growth in size leads to ground and excited state energy values observed in the bulk material (d). It is in the intermediate size range that size-dependent properties are observed. This phenomenon is discussed in more detail in Chapter 1, Nanotechnology and Chemistry: The Unseen Scale with Magnificent Impact.

For semiconductors such as Cu_2O, the collection of ground state energy levels containing valence electrons is called the valence band. The group of excited state energy levels is the conduction band due to the increase in electrical conductivity observed when an electron is excited from the valence band. The energy difference between the valence and conduction bands is referred to as the band gap. The band gap decreases as individual units of a semiconductor aggregate to form a nanoparticle and eventually a bulk (micron scale) particle.

A translucent solution's colour, its most obvious optical property, is due to the wavelengths of light that pass through it. The human eye detects the colours of light transmitted through the solution and does not detect the colours absorbed by the particles in the solution. As the energy difference between the ground and excited states decreases, the growing particle is capable of absorbing lower-energy photons, meaning that only the lowest energy photons at the red end of the visible spectrum will pass through the solution. As semiconductor nanoparticles increase in size, the solution colour changes from blue to green, yellow, orange, then red.

This visually appealing phenomenon is observed in many other semiconductors besides Cu_2O, including CdS,[12] CdSe,[13] and more complex structures.[14] The synthesis and analysis of optical properties of Cu_2O is presented here due to its use of less environmentally hazardous reactants, simpler preparation, and lower cost. It should be noted that ZnO and TiO_2 are also semiconductors but their band gap is sufficiently large that even nanoparticles absorb UV light but do not absorb any visible light.

Two methods for synthesizing Cu_2O are described. The first method starts with a copper(ii) salt that is reduced to Cu(i) oxide nanoparticles.[15] The second method also begins with copper(ii) but it is reduced to Cu nanoparticles, which then oxidize to copper(i) oxide.[16]

5.2.2.1 Size-dependent Colours of Copper(i) Oxide Particles

Students can synthesize Cu_2O by first dissolving copper(ii) sulphate in hot water, then adding sodium carbonate and sodium citrate. This solution is known as Benedict's reagent. It is stable for several weeks if stored in a sealed container. In a test tube, student mix Benedict's reagent with glucose and dilute with sodium hydroxide. The growth of Cu_2O nanoparticles is immediately visible when the test tube is placed in a beaker of boiling water. The reaction should continue for five minutes then it stops when the test tube is placed in a beaker of room temperature water.

Students follow the same procedure but vary the concentration of NaOH used to produce particles with diameters ranging from 170 nm (no NaOH) to

2300 nm (1 M NaOH). As expected, the smaller particles absorb only the higher energy photons of the visible spectrum, resulting in a faint blue coloured Cu_2O nanoparticle solution. The largest micron-sized particles have band gaps narrow enough to absorb all wavelengths of visible light except red, resulting in a reddish-brown coloured solution.

Students can also prepare Cu_2O particles with a narrower range of sizes (1.1 to 2.2 microns) by varying the amount of Benedict's reagent while keeping the sodium hydroxide concentration constant. This results in solutions of particles with colours ranging from yellowish-orange to red.

Due to the relatively large size of the nanoparticles (several hundred nanometres), the optical properties of these solutions are not as simple as is previously described by Figure 8.2. The particles are large enough to scatter light as well as absorb it, causing nanoparticle solution colours to change depending on the lighting. The full description of the optical properties is likely to be too complicated for a high school curriculum but, regardless, a complete understanding is unnecessary. The simple explanation of size-dependent colour provided here should be sufficient.

Students can directly observe the presence of at least the larger nanoparticles (<500 nm diameter) with a microscope commonly used in biology classes. Instructors can direct students' attention to the distribution of particle sizes within each sample and could even collect particle size distribution data using free software such as ImageJ.

5.2.2.2 Copper and Copper(I) Oxide Nanoparticles

This second preparation is distinctive because students prepare two nanomaterials.[18] First, they reduce copper(II) ions to copper metal nanoparticles. These are unstable and oxidize to Cu_2O. The experiment also demonstrates the greater reactivity of nanoparticles compared to their bulk analogues.

Like semiconductor nanoparticles, metal nanoparticles provide an example of size-dependent colour. However, the mechanism causing the colour change is different. Because valence electrons move easily through a metal, those electrons can collectively oscillate on the metal's surface. When the oscillation of the surface electrons couples with the electric field of light waves that strike the metal's surface, the electrons can absorb the light waves having a frequency equal to the electrons' collective oscillation frequency. This creates an additional absorption band (called a surface plasmon resonance or SPR band) in the absorbance spectrum of the particles, changing the colour of the solution. The frequency of the SPR band changes with the particle size. As we will see, the SPR band appears in the visible spectrum for copper, silver, and gold nanoparticles. (This phenomenon is discussed in more detail in Chapter 1, Nanotechnology and Chemistry: The Unseen Scale with Magnificent Impact.)

As will be described below, copper, silver, and gold nanoparticles are all easy to prepare. Copper nanoparticles have an advantage that they require less expensive starting materials. They are also less stable (more reactive) than silver or gold particles. This may prevent their use in experiments that require several days but the increased reactivity of the copper nanoparticles illustrates that particle size affects reactivity. This is a critically important aspect of nanomaterials that a synthesis of copper nanoparticles allows students to directly observe.

A simple but useful model describes nanoparticle stability. Atoms (or ionic units) are either on the surface or in the interior. Interior atoms are stabilized through bonds with other atoms or ions around them. Surface atoms lack at least some of these interactions since they are exposed on one or more sides. The exposure increases for atoms that are at a corner or edge, increasing their reactivity. Smaller particles have a greater portion of their atoms on the surface (due to a larger surface area to volume ratio) so they are more reactive. Larger particles have a relatively smaller number of atoms on the surface, leading to chemical properties that are more like bulk material properties.

Reduction of copper(II) occurs within minutes as students mix the reducing agent $NaBH_4$ in a solution of copper sulphate and iodide ions (present as a stabilizer). The presence of 10–20 nm diameter Cu nanoparticles is responsible for the colour change of the solution from transparent to a deep red. Within 30 min, the red colour changes to yellowish-orange as copper atoms and dissolved oxygen gas react to form Cu_2O nanoparticles. The colour of the Cu_2O nanoparticles is discussed in the previous experiment description. The loss of copper nanoparticles is reversed by adding an additional portion of borohydride that reduces copper(I). The original authors of this experiment note that the experiment can be completed in less than one hour and that the results are reproducible even when students add too much or too little of a reactant to the solution during preparation.

Metal nanoparticles are often stabilized by ions such as halides or citrate that bind to their surface. The presence of the anions creates an overall negative charge on the nanoparticle's surface. This in turn creates repulsion between particles, which lowers the likelihood of their collisions in solution and eventual aggregation. The repulsion between nanoparticles can be undermined by the presence of a high concentration of ions (*e.g.*, NaCl). The presence of non-reacting ions lowers the long-range electrostatic forces experienced by particles. Both the attraction between particles with opposite charges and repulsion between particles with the same charge decreases. Lowering the attractive or repulsive force affects the chances of a collision between the two particles, which then affects the rate of a reaction occurring. This is known as the kinetic salt effect. Although this topic is usually presented quantitatively in a college physical chemistry course, the author explains the qualitative aspects of the kinetic salt effect to high school students using the following analogy.

> "Consider that you and a person you do not like are invited to a party. You both arrive early. Neither wants to talk to the other so you both keep your

distance. This is the case for nanoparticles with negative charges on their surfaces – they experience electrostatic repulsion from each other so they stay away. (Unlike the people in the analogy, the nanoparticles are always in motion with the solvent and there is a chance that due to their random movements, the repelling particles do in fact collide.) When more people arrive, the room is more crowded and chaotic with everybody moving around, dancing, and carrying on. It is harder to keep tabs on your antagonist and occasionally you accidentally get close enough to exchange unpleasant words before moving away. This is the situation when some ions are present in solution. The non-reacting ions screen the ionic charge of each other and any other ions, including those on the surface of the nanoparticles. With less repulsion, a nanoparticle is more likely to approach another. At the height of the party, the room is packed. You can barely see beyond a few people directly in front of you. As you move, you turn and come face to face with the person you cannot stand to be around. At this point in the solution, the two nanoparticles combine. This process occurs many times, yielding a solution full of larger nanoparticles."

For nanoparticles with negative charges on their surface, the presence of excess ions in solution will lower the repulsion felt between them and increase the rate of aggregation. Nanoparticle size quickly increases.

As discussed earlier, the colour of metal nanoparticles will change due to a change in the position of the SPR band. The absorption energy of the SPR band depends on the particle's size. For aggregated copper nanoparticles, the SPR band shifts to higher wavelengths, creating solutions with a blue-green colour.

An interesting side note about the reaction: iodide ions are an effective stabilizer for Cu nanoparticles but other halides are not. The stability is due in part to the strong interactions between I^- (but not Cl^- or Br^-) and any Cu^+ ions that form on the copper surface. The greater attraction between Cu(I) and iodide ions is reflected in the K_{sp} value for CuI, which is many orders of magnitude smaller than K_{sp} for other Cu(I) halide compounds.

5.2.3 Gold Nanoparticles

While a first impression might be that gold is too expensive for a high school lab activity, this experiment uses a small volume of a low concentration gold salt solution. The experiment is easy to perform and provides vivid examples of nanoparticle properties.[17,18] Students can use the gold nanoparticles in several applications (see the lab activity described in Section 6.5 in Chapter 6).

In this experiment, students prepare gold nanoparticles in an aqueous solution and study the particles' behaviour when different solutes are added. Certain types of solutes cause the gold nanoparticles to come together and form bigger particles. Students then identify the property of the solute which leads to a colour change based on the results.

To make and study gold nanoparticles, students first reduce gold(III) ions to form metallic gold. The gold atoms will agglomerate together so this process must be stopped before the nanoparticles get too large and precipitate. Students accomplish both steps using an excess amount of citrate ions. Citrate is a reducing agent that will reduce Au(III) ions to elemental gold. As the gold metal atoms begin to bond together, more citrate ions will cover the surface of the gold, preventing the nanoparticles from growing larger than approximately 12 nm.

Students begin with 20 mL of a 1.0 mM solution of gold(III) stock solution ($HAuCl_4$) in an Erlenmeyer flask. This is diluted with water to approximately 50 mL. This initial solution is yellow. Students can mark the level of the solution with a felt tip pen on the outside of the flask. The flask is heated to boiling and stirred. As water evaporates, additional water is added to maintain the solution volume. At this point, students add 2 mL of a sodium citrate solution to the flask. This causes a colour change as the nanoparticles form, first from yellow to clear, dark blue, and finally to a deep red wine colour. Students should continue to heat and stir for a few more minutes to ensure that the reaction is complete. Students can then remove the solution from the hotplate to cool. The dark red colour is due to the presence of gold nanoparticles. After the solution has cooled, Students should be able to see through the solution in the flask. if it is too dark, they can dilute it with perhaps an additional 10–20 mL of water.

An alternative preparation uses a solution of gold(III) ions reduced by a commercially available green tea extract.[19] The reaction takes about the same amount of time as the citrate reduction and does not require boiling the solution. Before proceeding to the next part of the experiment (which the author has performed with the citrate-reduced Au nanoparticles but not those prepared with the tea extract), students will need to filter the gold nanoparticle solution to remove green tea solids from the tea extract.

In the next step, students use the gold nanoparticles as a rudimentary sensor. They should add a few millilitres of the gold nanoparticle solution to a series of test tubes. The instructor can provide a series of aqueous solutions that students will add to the test tubes. Suggestions include NaCl, $Ca(NO_3)_2$, HCl, NaOH, acetic acid, glycerol, $Na_3C_6H_5O_7$ (sodium citrate), and methanol. The selection of solutions should include both ionic and molecular compounds and all concentrations should be approximately 1 M. Strong acids, strong bases, and neutral ionic compounds will cause the gold nanoparticle to aggregate, changing the solution from a wine-red colour to a grey-blue colour, due to the kinetic salt effect described in Section 5.2.2.2. Optical spectra of the two solutions are shown in Figure 5.2. Adding molecular compounds to the gold nanoparticle solutions will not cause a colour change. Weak acids or bases like acetic acid will cause a small colour change due to incomplete dissociation.

It can be helpful for students to share their data, perhaps by tallying the results in a table on the board. In this way, students (and the instructor) can identify any results that are outliers. Working in groups, students can

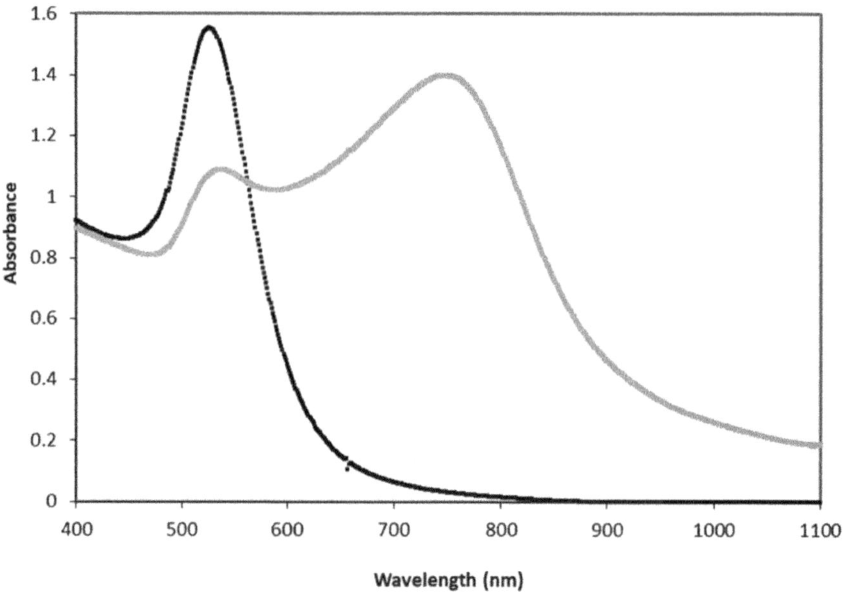

Figure 5.2 Spectra show the absorbance of solutions containing separate gold nanoparticles (red wine colour, black line) and aggregated nanoparticles (grey-blue colour, grey line).

usually identify the feature of the solute (formation of ions in solution) that causes a colour change. This colour change shows how a simple gold nanoparticle solution can act as a sensor for ionic compounds.

The citrate reduction of gold(III) procedure yields more solution than students need for the exploration of the Au nanoparticle sensing properties. It is possible to begin with less than 20 mL of the initial 1.0 mM gold(III) solution. However, if students do have excess solution left over, they can use it to make a moulded plastic shape or colour in a picture.[20] They add a few drops of polyvinyl pyrrolidone (PVP) solution to the nanoparticle solution to prevent particle growth. Note that students should not add PVP to the entire batch of gold nanoparticles since PVP will prevent the colour change desired in the sensing activity. They heat the nanoparticle solution until it is almost boiling. To this, they slowly add a small amount of polyvinyl alcohol (PVA) and stir. The solution should then boil until all the PVA dissolves. Concurrently, students can lightly grease the sides and bottom of a mould with silicone grease. Once the PVA has dissolved, it can be removed from the hot plate and allowed to cool. Students should gently pour the nanoparticle solution into the mould, avoiding the formation of bubbles if possible. They carefully move the mould to a toaster oven preheated to 150 °C and allow the water to evaporate (90 min). Once removed from the oven, students can gently peel the plastic (PVA) sheet with embedded nanoparticles from the mould. The material is safe to handle and take home. Silver nanoparticles can also be used for this activity.

5.2.4 Silver Nanoparticles

Like copper and gold, silver nanoparticles form easily and display a range of colours, from yellow to grey. Educators have developed many experiments for preparing and using silver nanoparticles because the starting material (typically silver nitrate) is cheaper than gold salts and silver nanoparticles are more stable than copper nanoparticles. The reduction of silver cations to form silver nanoparticles is possible using a wide variety of reducing agents, including naturally occurring materials. This expands the potential learning objectives to include a better understanding of environmentally benign (green) chemistry techniques. Students can use their silver nanoparticles to learn about antimicrobial agents, chemical toxicity, and a method for removing nanoparticles from the environment. Depending on the synthesis method, silver nanoparticles can be stable for months if kept in a brown glass or opaque container. Silver is photo-reactive so clear glass containers should be avoided for storage.

5.2.4.1 Silver Nanoparticles from Inorganic Reducing Agents

Chemists use the Tollens reaction to selectively indicate the presence of aldehydes. In a basic solution, an aldehyde is oxidized by $Ag(NH_3)_2^+$ (silver diamine cation) to form a carboxylic acid and the silver cation is reduced to silver metal, which adheres to the inside surface of the glass container. Other oxygen-containing functional groups such as alcohols and ketones do not participate in this reaction. By modifying the concentrations of each NaOH, silver diamine, and glucose (a saccharide containing an aldehyde group), students can perform the Tollens reaction to create bulk, submicron-sized, and nanosized silver. Each has a characteristic appearance.[21]

A coating of large silver particles forms the reflective surface of a mirror. This is prepared by first pre-treating a clean glass surface (*e.g.*, a microscope slide) with sodium bicarbonate and tin(II) chloride solutions. Next separate solutions, one containing silver nitrate and ammonium nitrate, and the other containing D-glucose, tartaric acid, and sodium hydroxide are mixed. When the microscope slide is immersed in this combined solution, silver particles deposit on the surface within ten minutes. The outward silver surface is not smooth so it is not a good mirror. Scraping off the silver layer from one side of the slide exposes the smooth, reflective layer of silver adhered to the other side of the microscope glass.

Submicron-sized particles of silver scatter all wavelengths of visible light so they appear black. Such silver particles are the source of the black colour in black and white photography. Students can mix $AgNO_3$, NH_3, NaOH, and D-glucose in lower concentrations compared to those used in the preparation of a mirrored surface. Large silver particles form due to the Tollens reaction.

Another slight variation in concentrations of the same reactants leads to silver nanoparticles growing in solution. Initially, the smaller nanoparticles absorb light below 500 nm, giving the solution a yellow colour. The particles grow to 40 nm in diameter, shifting the SPR band to 450 nm, and changing the solution colour to orange.

Smaller silver nanoparticles form through the reaction of silver cations with borohydride (BH_4^-).[22] An excess of borohydride is used because it serves as both a reducing agent and stabilizer for the nanoparticle surface. Students add a silver nitrate solution dropwise into a stirring beaker of aqueous sodium borohydride that is sitting in an ice bath. The dropwise addition of $AgNO_3$ is conveniently performed by dispensing the silver nitrate solution from a buret with the stopcock only slightly open. Silver nanoparticles with a diameter of about 12 nm form, creating a bright yellow coloured solution. Stirring should stop at this point but if it continues, the agitation will cause the particles to continue to grow. Again, the borohydride is playing a role in the system, this time by providing non-reacting ions that promote aggregation *via* the kinetic salt effect. Nanoparticle growth causes a deeper yellow colour, followed by a change to orange, lavender, then grey. The orange and grey colours are reminiscent of the solutions formed in the Tollens synthesis of Ag nanoparticles described earlier.

5.2.4.2 Silver Nanoparticles Formed Through Green Chemistry

Green chemistry is a philosophy and experimental approach to performing chemistry that protects chemical workers, product consumers, the public, and the environment. Like nanotechnology, green chemistry provides new ways to think about chemical reactions and materials that students should understand. Green chemistry is defined by the following principles.[23]

Prevent waste by designing reactions to not produce waste. Use as many atoms as possible from the starting material when making the product. This is known as "maximizing the atom economy" and addresses waste production from the perspective of the atomic scale. Reducing waste in reactions includes avoiding the use of temporary protecting groups that must be removed later. Catalysed reactions are preferred because the catalyst can be reused. Consider the waste produced by products that have exceeded their useful lifetime. Produce materials and chemicals that degrade after they are no longer needed.

Use and produce chemicals that are safe. Protect people and the environment by designing reactions that involve nontoxic materials. Design processes that avoid the occurrence of accidental explosions, fires, or release of chemicals into the environment. Continuously monitor processes to prevent pollution or waste.

Reduce energy use. Design reactions to occur at room temperature and ordinary pressure. Catalysts can also increase energy efficiency of a reaction. Design reactions to use renewable starting materials, such as waste materials from other chemical or agricultural processes.

Nanotechnology can be consistent with green chemistry principles. By building materials at the atomic scale, it is possible to have greater control over product formation, reducing waste. Due to the variety of size-dependent properties at the nanoscale, scientists can now choose from more materials for a desired application. Smaller devices, such as computer chips, require less energy to function. With these green chemistry principles in mind, educators have designed many experiments for preparing silver nanoparticles.

A quick synthesis is available by heating solutions of D-glucose and silver nitrate in a household microwave.[24] This process requires much less energy and fewer reactants to prepare Ag nanoparticles compared to other methods. Short heating times produce brown solutions of silver nanoparticles with an absorbance peak that spans the entire visible spectrum. This author finds that the brown solution colour is not as visually appealing as the hues produced by other syntheses. Due to the variability of power delivered by an inexpensive household microwave, the extent of the formation reaction is inconsistent among repeated trials. Still, if time or reagent budget is limited, this is an adequate approach for making silver nanoparticles.

A second microwave-assisted synthesis method is available and claims to provide more reliable results.[25] This is likely due to the use of starch and citrate, which can stabilize the nanoparticles and prevent excessive growth. The absence of starch from the procedure causes the nanoparticles to be generally less stable; nanoparticles formed without starch are sensitive to the addition of non-reacting ionic compounds. In the presence of starch, the nanoparticles are more stable and unaffected by the addition of NaCl. This is not surprising since starch is a molecular compound so there are no electrostatic forces between nanoparticles that added ions could affect. D-Glucose is used as a reducing agent for silver nitrate. Nanoparticles have a diameter of 5 nm. Students can use a hotplate to prepare the nanoparticles as well. This experiment was performed by undergraduate students but it appears to be sufficiently straightforward for high school students.

Students can use a variety of sugars to form silver nanoparticles, including D-fructose, D-xylose, and D-mannose.[26] Interestingly, the reduction of silver cations by D-fructose takes much longer (8 min) in a microwave while other monosaccharides require just one minute of microwave heating. Particles prepared using any of these reductants are about 25 nm in diameter. All solutions are a dark yellow colour and become more reddish brown if heating continues due to the continued growth of the nanoparticles.

Naturally sourced reductants are capable of preparing Ag nanoparticles. For instance, students can add silver nitrate to a solution of diluted honey, heat to 80 °C for 10 min, then stir for an additional 30 min.[19] Unlike the microwave heating method, this procedure reproducibly yields a bright yellow coloured solution of smaller Ag nanoparticles. This synthesis was part of a larger, inquiry-based activity in which students reviewed the scientific literature and developed their own procedures.

If an instructor prefers a less sweet reducing agent, onion extract can react with silver nitrate to produce brown nanoparticles of silver.[27] Students boil diced onion then filter the mixture to remove solids. This is mixed with a high concentration (0.1 M) of silver nitrate. The original experiment uses the silver nanoparticles to enhance the electrochemical degradation of pollutants for an instrumental analysis laboratory course.

Plant leaves contain proteins and carbohydrates that can serve as reducing agents also. This is demonstrated using geranium leaves that have been chopped, boiled, and centrifuged.[28] Students add the leaf extract to a silver nitrate solution then heat below boiling for 20 min. The solutions have a yellow or brown colour with a broad absorbance peak. The authors note that they tried other sources of leaf extracts with mixed results.

5.2.4.3 Applications of Silver Nanoparticles

Antimicrobial properties of silver nanoparticles are well known and their use in commercial products is increasingly common. However, the toxicity of silver nanoparticles is not limited to harmful microbes. Silver negatively impacts plants and animals also. One method of removing silver and other nanoparticles from water and soil is phytoremediation, the absorption of a pollutant by plants which are then harvested. Students should learn about the benefits and negative impacts of nanotechnology so that they can develop their own balanced view. The following activities address biological and environmental issues and demonstrate the interdisciplinary nature of nanotechnology.

Instructors should consider whether to include the synthesis of nanomaterials within an overall activity that includes their applications. Instead of synthesizing nanoparticles, instructors could purchase them. This might be an appropriate option if suitable laboratory facilities and supplies are unavailable, there is limited time to perform the activity, or students have not developed the necessary laboratory skills needed to perform a synthesis. Another advantage is that the purchased nanoparticles are more likely to yield reproducible results, which ensures that all students will learn more about the application. Students who fail to properly synthesize nanoparticles in the first part of an activity will not be able to use them for a subsequent application. Instructors should note that size, coating, and shape can affect the properties of nanoparticles. A reputable manufacturer selling nanoparticles should be able to recommend a particular product for a given application. During that discussion, instructors might ask for a free sample, which vendors are often happy to provide since it will be used in an educational setting.

Untreated wastewater is a major health concern in underdeveloped countries. Silver nanoparticles are toxic to microbial organisms (bacteria, viruses, fungi, *etc.*) in wastewater. Students can observe the effect of nanoparticles on the growth of microorganisms using a simple microscope in water samples collected by the students themselves.[29] Students collect wastewater samples from their homes or other familiar locations. They add portions of their water sample to separate, sterile petri dishes then add nanoparticle solution to one. The other portion of wastewater is the control. The water samples incubate for several hours. Students pour samples of water from each dish on agar in a petri dish then incubate them for 24 hours. Microorganism growth is obvious in the sample that did not contain added Ag nanoparticles. Treated wastewater samples showed less or no microbe growth. Students can identify the type of microbe by examining them under a microscope.

This experiment is designed for junior high school students and uses commercially available Ag nanoparticles. It could be modified for high school students by allowing students to synthesize their own nanoparticles using one of the previously described methods. The instructor should verify that a particular synthesis method produces nanoparticles that are toxic to microbes. Note that some reducing agents such as borohydride form oxidation products that are themselves toxic. Proper control reactions ensure that the toxicity is due in fact to the nanoparticles. Another possible change is to prepare different concentrations of nanoparticle solutions to determine the minimum concentration necessary to prevent microbe growth.

Students can study the nanoparticle toxicity towards other organisms. In this experiment, students prepare gold or silver nanoparticles using a citrate reduction method (gold) or combination of citrate and borohydride (silver).[30] They perform serial dilutions which are mixed with solutions of brine shrimp in well plates. Students count the number of living brine shrimp before and after 24 hours of exposure to nanoparticles. The authors point out that counting the tiny brine shrimp is difficult so students must take their time, otherwise there can be significant error in their measurements. A class of students can combine their data to graph and perform a statistical analysis.

A simplified version of an experiment designed for undergraduate students in a first-year chemistry lab[31] can help students learn about the effects of nanoparticles on plants. In the original experiment, students synthesized Ag nanoparticles using a citrate reduction in hot water, like the one described above.[25] The procedure is described in ref. 31 but the primary source's description, developed by the California NanoSystems Institute, is no longer available online. In the next step, students added some of the nanoparticle solution to test tubes containing stalks of water weeds such as *Egeria densa*. After a week of exposure, the plant leaves showed a dose-dependent loss of green colour. This was confirmed during the third part of the experiment in which students extracted chlorophyll from the plant leaves and measuring its concentration. High school students can complete a similar experiment but without the third step (chlorophyll extraction). Instead, they can perform a visual analysis of the plant leaves.

5.2.5 Solar Cells

Efficient conversion of solar energy to electricity is a means for scientists and engineers to achieve the goal of reducing the use of fossil fuels. The most recent generation of solar cells use titanium dioxide nanoparticles and organic dyes or organic films containing nanostructures. These solar cells are less expensive manufacture and use more environmentally benign materials. Different published activities enable advanced high school students to build and test their own dye-sensitized solar cells. Addressing climate change is an important issue especially to students so these activities should interest them. Due to the complexity of the procedure and the number of steps involved, instructors should carefully consider how much assistance to provide to their students for them to successfully complete the activity. From another perspective, the additional challenges of these activities allow instructors to incorporate concepts of electricity and engineering. Unlike previously discussed activities, students actually *build* a device that relies on chemistry for its operation.

In the first procedure,[32] students spread a layer of paste containing TiO_2 nanoparticles on a conducting glass slide. This is the cell's anode. Students then place the anode in a solution containing an organic dye that absorbs in the visible spectrum. The cathode is prepared by depositing soot from a candle on another conducting glass slide. The anode and cathode slides are combined so that the TiO_2 paste and soot layers are in contact with each other and secured with binder clips. A solution of potassium triiodide acting as the redox electrolyte is injected along the edge and diffuses between the two slides. Students connect alligator clips to the anode and cathode slides during irradiation of the solar cell to generate a current and voltage. This is measured by a multimeter. A class of students can connect their individual solar cells in series (Figure 5.3) to increase the total voltage or in parallel to increase the total current.

Figure 5.3 Dye-sensitized solar cells are connected in series. Students measure the current and voltage with a multimeter.

There are various ways to prepare the ingredients for this activity. A powder of TiO_2 nanoparticles is available commercially, which can be mixed with dish soap and acetic acid to make the paste.[32] Titanium dioxide nanoparticles are also prepared by calcining sunscreens[10] (see above), toothpaste, or foods such as powdered sugar.[33] The light-absorbing dye can be from many types of fruit juices.

A more sophisticated version of this procedure requires the purchase of more commercial chemicals and materials.[34] The solar cells are capable of powering LEDs and a small fan. The authors attribute students' strong interest in this activity to their ability to see the solar cell performing a practical task, rather than reading current and voltage from a multimeter.

A third procedure involves the use of organic films of less than 250 nm thickness to convert light to electricity (organic photovoltaics or OPV).[35] This procedure also uses an ink containing nanosized carbon particles called fullerenes. The authors describe this new technology in detail within the original publication and provide thorough instructions for preparing solutions and for 3D printing an inexpensive spin coater that students use to deposit the organic thin films. The design of the spin coating device allows students to vary the film thicknesses and observe their effects on the efficiency of the photocell. Thicker films absorb more light that can yield the necessary charged particles but thinner films reduce recombination of photogenerated charged particles. Like the previously described solar cell description, this procedure requires materials and chemicals that are not common for a high school laboratory.

5.2.6 Magnetic Fe_3O_4 Nanoparticles

Ferrofluids are a fun and visually impressive demonstration of nanotechnology. A ferrofluid contains nanoparticles that respond to a magnetic field suspended in a liquid. The nanoparticles are made of a ferrimagnetic material such as magnetite (Fe_3O_4). Since the nanoparticles are dispersed within the solvent, the nanoparticles' interaction with the magnetic field causes the entire liquid to respond. When holding a magnet near a ferrofluid, students can easily observe some stunning behaviours. There is a significant increase in the viscosity of the liquid, the liquid moves as it follows the movement of the magnet nearby, and the magnetic field causes a distortion of the liquid's surface with the formation of spikes as the nanoparticles follow the magnetic field lines of the magnet. Figure 5.4 shows these spikes when a magnet is held underneath the plastic dish of ferrofluid. This is analogous to the positioning of iron filings when a magnet is nearby (see the lab activity described in Section 6.6.1, Chapter 6).

Ferrimagnetism exists when the overall spin of electrons within the crystal structure is non-random and are aligned with an external magnetic field. Nanoparticles remain suspended in a liquid, forming a colloid, due to their strong interactions with the solvent. When a magnetic field attracts the ferrimagnetic nanoparticles, the solvent is pulled along. One might think

Figure 5.4 Spikes form in a ferrofluid when a magnet is held underneath the dish.

that the way to create a liquid that responds to a magnet is to heat a ferrimagnetic solid and melt it. This would not work because the heat will provide enough energy to randomize the orientation of the electron spins before the solid melted. Materials with random electron spins do not respond to a magnetic field.

All activities described in this section rely on ferrofluids containing magnetite nanoparticles. Students can easily synthesize this as in aqueous solution. Water from an aqueous ferrofluid will evaporate within a few hours so their long-term storage is very difficult. Oil-based ferrofluids are more stable and interact more strongly with a magnet. Their synthesis is beyond the scope of a high school or first-year undergraduate chemistry course. They are available commercially and are appropriate for class demonstrations.

Instructors should understand and convey to students some guidelines about working with ferrofluids. Ferrofluids are messy. The magnetite stains clothes and skin and the aqueous solution has a high pH. A ferrofluid liquid is difficult to remove from magnets. There should always be a material such as a plastic dish separating the magnet and ferrofluid. Students will be handling neodymium magnets, which are much stronger than common refrigerator magnets. The magnets can be difficult to pry off from a smooth metal surface and they will damage magnetic media. While cassette tapes and floppy disks are almost extinct now, credit cards and ID cards do contain data stored in their magnetic strips. The magnets will quickly slide towards other magnets that are several inches away.

Beyond synthesizing a ferrofluid, there are many activities that students can perform that demonstrate its applications. Activities are presented in order of increasing difficulty.

The synthesis of a ferrofluid is straightforward and is suitable for middle and high school students.[36] The chapter author provides recommendations here based on performing this experiment many times. Iron(II) chloride solid slowly oxidizes to Fe(III) over time, changing from light green crystals to brown. It is advisable to purchase smaller containers of $FeCl_2$ more frequently, rather than one large batch. An $FeCl_2$ solution will turn brown within hours of its preparation. Students can dispense the iron chloride solutions using disposable plastic syringes (with the needle removed) more quickly than using graduated cylinders. The syringe volume markings are sufficiently accurate. The syringes can be reused until their volume markings rub off from use. Solutions can be stirred by hand or with a stir plate. If using the latter method, students should stir as slow as possible at first to avoid splashing. Household ammonia that does not contain any surfactant is suitable for use when NH_4OH is needed. The ferrofluid initially formed contains 15 nm diameter Fe_3O_4 nanoparticles along with other by-products and unreacted chemicals also. To isolate the ferrofluid, students place a magnet next to the beaker or cup containing the ferrofluid. Within a few minutes, all the ferrofluid collects close to the magnet. While still holding the magnet in place on the outside of the container, students decant the unneeded material. Some black solids resembling the desired ferrofluid may be poured out. Students should only keep the ferrofluid that is attracted to the magnet. The tetramethyl ammonium hydroxide surfactant liquid degrades over the span of years, producing a strong fish odour. The final step in the ferrofluid synthesis is to remove as much excess water as possible. By carefully holding the magnet underneath the ferrofluid dish, students can gently shake a drop or two of water out of the ferrofluid if necessary. Moving the magnet back and forth causes the ferrofluid to move, which seems to promote water evaporation that can help concentrate the Fe_3O_4 in the solution. Holding the magnet close to the underside of the plastic dish, rather than touching it, maximizes the height of the spikes that form. The formation of spikes indicates that the ferrofluid synthesis is complete and successful.

Once students prepare their ferrofluid, they can begin investigating some of its properties and uses. With a toothpick or stirring rod, students can transfer a drop of ferrofluid to a small piece of paper towel. After it dries, the nanoparticles act as a magnetic ink, similar to an anti-counterfeit measure used in paper currency. A magnet can lift the piece of paper that has soaked in the ferrofluid.[36]

Students can compare their ferrofluids by placing squares of paper between a neodymium disk magnet and the underside of the ferrofluid dish. As the separation between the magnet and ferrofluid increases, the attraction between them decreases. At a certain distance, the attractive force is less than the gravitational force pulling down on the magnet. Samples of ferrofluids prepared properly and recovered in the greatest amount will require the most pieces of paper to cause the magnet to drop.

Other activities include preparing ferrofluid slime (crosslinked PVA with ferrofluid), a ferrofluid plastic, and a ferrofluid powder.[37] The slime and plastic are especially fun for students because they can handle the ferrofluid-containing polymer materials, whereas they could not touch the ferrofluid for fear of staining their skin. Preparing the powder allows for safe, long-term use of the ferrimagnetic nanoparticles.

Preparing a magnetic ink is not the only real-world application of ferrofluids. If students add activated carbon powder to the initial solution of iron chloride salts, they can prepare the ferrofluid with nanoparticles embedded in the carbon material.[38] Activated carbon is used for removal of organic pollutants from wastewater but the powder is difficult to recover. By including ferrimagnetic nanoparticles within the powder, it can be separated from a solution using a magnet. Adding the ferrofluid-activated carbon composite to a solution of organic dye results in adsorption of the dye, which is visually confirmed. Students can separate the ferrofluid-activated carbon material from the solution with the use of a magnet and disperse it in ethanol, which causes the dye to desorb. The ferrofluid-activated carbon composite is reusable in this manner. This application was performed by first-year college students but the use of activated carbon does not appear to introduce any tasks that a high school class cannot perform.

Two published reports describe similar uses of ferrofluids to remove arsenic[39] and metal cations from water.[40] Both require equipment such as fume hoods, reflux condensers, and vacuum ovens that are not common in high schools. The instructor prepares the ferrofluid for these activities, using a different method than the one used in previous activities. The instructor also performs the final detection of heavy metal cations[40] but a modification to the procedure would allow students to qualitatively identify the presence or absence of metal ions.

Ferrofluids have many useful applications but perhaps the most interesting is as a medium for sculpture. Artist Sachiko Kodama uses ferrofluid to create kinetic sculptures.[41] By varying the magnetic field, the ferrofluid flows, spikes, and moves on a surface. An episode of Science Tuesdays demonstrates how ferrofluid flows over a magnetized metal surface in a manner like the Sachiko Kodama artwork.[42] Artist Dakd Jung built a speaker that contains an electromagnet that causes ferrofluid to dance to music.[43]

5.2.7 Metal Oxide Nanothin Films

A thin film, such as oil on water or soap bubbles, can produce a pattern of colours due to thin film interference. The light waves reflecting off the top surface layer interfere constructively or destructively with the light waves that pass through the top layer, through the lower layer, then reflect upward through the top layer. The film will have a nanoscale thickness to produce a pattern of yellow, blue, and purple colours. A solid thin film of a metal oxide can produce a similar effect.

As a class demonstration, an instructor can produce coloured metal oxide films on the surface of metals that are heated on a hotplate or in a propane flame.[44] The colour changes occur during heating and persist once the material cools, so instructors could prepare the films ahead of time and reuse them as needed. Stainless steel and titanium nitride form metal oxide films with thicknesses of less than 100 nm. The metallic material does not have to be high purity. Tools such as titanium nitride coated drill bits are sufficient for this demonstration.

5.3 Hands-on Activities Using Models

For reasons of cost, safety, available time, and specific instructional goals, instructors may wish to engage their students with models to teach nanoscale phenomena. Still, activities using models provide opportunities for hands-on learning.

5.3.1 Scanning Probe Microscopy

Scanning Probe microscopy includes a collection of instrumental techniques for characterising nanoscale features of surfaces. The tip of a probe passes along a material's surface. Interactions between atoms of the probe's tip and the atoms of the surface are detected and recorded, creating a map of the surface. The type of interaction depends on the microscopy method. Two of the most well-known methods are scanning tunnelling microscopy (STM) and atomic force microscopy (AFM). Education-grade versions of these instruments are available and currently cost approximately $10 000 and $25 000 (USD) respectively. Manufacturers provide pre-made samples or users can prepare their own. Middle and high school students can gain a better understanding of the particulate nature of matter by using scanning probe microscopes under close supervision[45] or remotely.[46] Examples of STM and AFM images collected by students using education-grade instrumentation are shown in Figure 5.5.

One of the challenges with STM and AFM is that the tip must be as sharp as possible. If the tip is dull, it can interact with the surface in multiple points or not interact with atoms positioned in crevices of the surface. This limits the quality of the microscopy image. To better understand how tip sharpness affects image resolution, students are asked to examine the three-dimensional features of a surface by touch rather than looking at it. Surface features can be created using Lego® bricks attached to a base plate or wooden blocks glued to a paper plate. A student describes the surface by making successive "scans" while wearing mittens, heavy gloves, thick rubber gloves, with bare hands, then finally viewing the surface. This demonstrates the use of increasingly fine-tipped probes for imaging. Unless the student is visually impaired, sight will provide the ability to detect the most minute surface features, such as wood grain or chips.

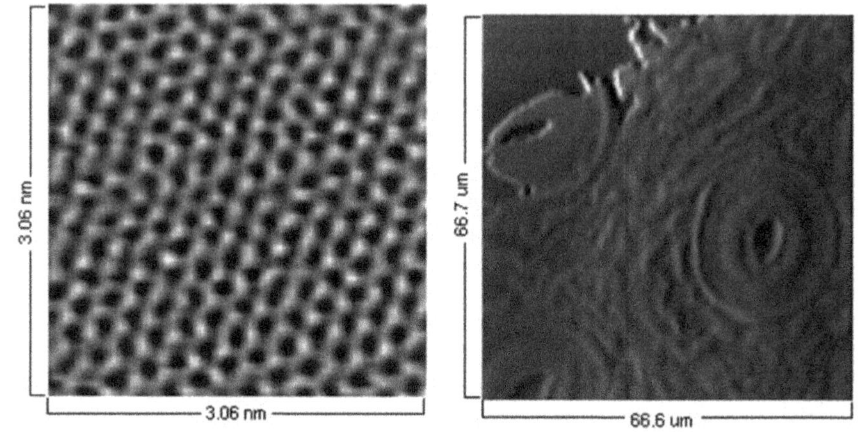

Figure 5.5 An STM image of highly oriented pyrolytic graphite (left) and an AFM image of a leaf stoma, both images collected high school students.

Imaging a surface with STM requires that the material be electrically conductive. This is because the instrument measures the very small amount of current that flows across the gap between the atoms of the tip and the surface atoms. Electrons move between the tip and the surface without the two making physical contact due to a quantum mechanical phenomenon called tunnelling. In violation of its behaviour predicted by classical physics, an electron has a non-zero probability of passing through a barrier such as the gap between the tip and surface. The tunnelling current changes according to the distance between the tip and the nearest surface atom. Moving the STM tip across a surface creates a map of the surface features. Using an education-grade STM, it is relatively easy to image graphite, gold, polymer strands, and nanoparticles (including ones described in Section 5.2). Under optimal conditions, students can image a surface with atomic resolution.

If an STM instrument is unavailable, students can use a simple model that illustrates important concepts about STM. Instructors can augment this experience by showing STM images that are available online. For this activity, students map a metal surface partially covered with masking tape using multimeter probes. This mimics the scanning of a tip across a conductive surface. Exposed areas are conductive, while spots covered by tape are not. Covering the metal sheet with fabric prevents students from knowing where the tape is until they scan the surface. One student positions a multimeter probe on an exposed part of the metal sheet and pokes the other probe through the fabric cover so that it contacts the surface. The multimeter will register a current if the probe touches metal. The student moves the second probe to a neighbouring spot to continue the scan. Another student can record the current at each location. Two students demonstrate this in Figure 5.6.

Figure 5.6 Two students measure the conductivity of different spots of a metal plate with multimeter probes, re-enacting the scanning process of an STM.

Atomic force microscopy relies on the deflection of a cantilever as the attached AFM tip comes in close contact with atoms of a surface. As the atoms of the AFM tip encounter atoms on the surface, the interaction causes the cantilever to bend. A laser shines on the cantilever arm and the location of the reflected laser light changes due to the degree of bending in the cantilever. A detector measures the displacement of the reflected laser light and converts that into a measurement of surface height. While the resolution of an AFM image is typically lower than that of an STM, AFM has several advantages. It is a more versatile method since the surface does not have to be conductive to collect an image. An STM image reveals an area covering about 100 nm^2 while AFM shows many orders of magnitude more. This is often desirable since optical microscopes allow a viewer to see surface features down to the micron level. Different types of AFM instruments allow users to detect other properties of the surface, such as magnetism, friction, and elasticity. AFM can also image surfaces that are submerged in a liquid. As with STM, students could collect AFM images if an instrument is available. Several hands-on activities simulate the use of an AFM.

Students can use any object with a narrow tip such as a wooden dowel or upside-down plastic bottle to simulate the AFM probe. The surface can be made from Lego® bricks, an egg carton, or other stacked objects.[47,48] Placing the probe at each point as it moves across each row of the sample generates surface height data which students record and graph, creating a three-dimensional image of the surface. The way students use their probe to contact the surface can reflect real modes of AFM imaging. In contact mode, the

AFM tip remains in contact with the surface as it moves. This can either damage the tip if the sample is very hard or damage the surface if the surface is soft. Since the tip is always in contact with the surface, there is a greater chance for atoms from the surface to attach themselves to the tip, reducing imaging resolution. A preferred mode is tapping, in which the tip touches the surface to record a measurement then retreats before it moves to the next location. Using a surface with clay features can illustrate to students that an AFM measurement can leave a lasting mark on the sample.

Students can build a simulated AFM that detects surface features using the same general method as a real AFM.[49] In this case, a laser pointer serves as a laser light source. Again, Lego® bricks are useful for building this apparatus, although other simple materials are suitable also. Students' AFM construction should include a tip connected to a cantilever which deflects light from a laser pointer as the tip moves across a surface. The reflected light should strike a surface that would allow the measurement of its deflection. Beyond that, students can express their creativity in the overall design of their AFM model.

A final AFM activity demonstrates magnetic force microscopy, in which a magnetized tip detects the presence of a magnetic field on the surface.[50] Common refrigerator magnets contain strips of the magnetic material with alternating orientations of north and south poles. A thin piece cut from one end of the magnet acts as a probe that students can slide across the rest of the magnet's surface. Students can feel the probe move as it is attracted and repulsed by the surface of the magnet, just as a real magnetic force microscopy tip interacts with a sample surface.

5.3.2 Models for Other Nanoscale Concepts

Two activities use models to demonstrate concepts that govern the behaviour of materials at the nanoscale: the surface to volume ratio of a particle and self-assembly of particles. Other activities that illustrate nanotechnology concepts are presented in Chapter 6, Nanochemistry as a Relevant Concept in Teaching Chemistry.

Instructors can stack small cubes together into a larger single cube to illustrate surface to volume ratio. While this can help students understand the basic concept, the use of nanoblocks shows this as well as the practical use of nanosized particles.[51] Nanoblocks are cubes covered in hook-and-loop (*e.g.*, Velcro®) tape. Students work with one large nanoblock and eight smaller nanoblocks that have a total volume equal to that of the larger cube. Students roll the nanoblocks in a tray containing small balls of fabric that stick to the nanoblocks. By counting the number of balls that stick to the nanoblock surface, students can measure the consequences of a collection of smaller particles having a relatively larger total surface area. This activity also models the behaviour of molecules that are adsorbed to the surface of nanoparticles for applications such as removal of air and water pollutants. Not only are more fabric balls (pollutant particles) removed by the smaller nanoblocks, but the adhesion of the fabric balls is also reversible, allowing the reuse of nanoblocks once they are cleaned. Students can even calculate an adsorption isotherm of fabric ball adhesion to the nanoblocks (more details are described in Chapter 1).

Particles often spontaneously organize themselves in water because attractive forces that exist amongst the particles and/or amongst water molecules are greater than the attractive forces that exist between the particles and water. This self-assembly leads to the spontaneous formation of structures at the nanoscale, such as micelles in mixtures of oil and water. Soda straws demonstrate self-assembly in water by orienting themselves side by side on the water surface.[52] Using straws of different length or diameter can demonstrate the complexity of self-assembly involving molecules with different shapes and sizes.

5.4 Conclusions

During the past several decades, nanotechnology has increasingly become a part of everyday life through research and development of new products and through the creation of hands-on activities for high school and middle school students. These activities include experiments in which students synthesize nanomaterials, explore their properties, and learn about their applications. They can manipulate model systems to better understand technologies such as scanning probe microscopy. Educators from around the world have contributed to this effort to prepare students for studying nanotechnology at the college level and pursue nanotechnology careers. This chapter highlights just some of the many activities available. Interested educators are encouraged to seek out additional published activities online and even develop their own so that this growth of nanotechnology education in the middle and high school grades can continue. As the field of nanotechnology grows and develops, more work should be focused on understanding students' misconceptions and conceptual difficulties. With the exception of ref. 1, the activities described here do not explore this important aspect of teaching and learning.

Acknowledgements

The author thanks Drs Jim Brenner and Joel Olson for their help leading many of the experiments presented here.

References

1. S. Y. Stevens, L. M. Sutherland and J. S. Krajcik, *The Big Ideas of Nanoscale Science and Engineering*, NSTA press, Arlington, Virginia, 2009.
2. G. M. Bodner, *J. Chem. Educ.*, 1986, **63**, 873.
3. T. W. Shiland, *J. Chem. Educ.*, 1999, **76**, 107.
4. L. Springer, M. E. Stanne and S. S. Donovan, *Rev. Educ. Res.*, 1999, **69**, 21.

5. G. Fischer *Proceedings of the Eighth Conference on Participatory Design: Artful Integration: Interweaving Media, Materials and Practices*, ACM, 2004, vol. 1.
6. W. J. Guedens, M. Reynders, H. Van den Rul, K. Elen, A. Hardy and M. K. Van Bael, *J. Chem. Educ.*, 2014, **91**, 259.
7. C. Walters, A. Keeney, C. T. Wigal, C. R. Johnston and R. D. Cornelius, *J. Chem. Educ.*, 1997, **74**, 99.
8. J. R. Abney and B. A. Scalettar, *J. Chem. Educ.*, 1998, **75**, 757.
9. H. P. Moeur, A. Zanella and T. Poon, *J. Chem. Educ.*, 2006, **83**, 769.
10. J. Dege, T. Waitz, S. Haffer, V. Pietzner, R. Abu-Much, M. Hugerat and T. Wilke, *J. Nano Educ.*, 2016, **8**, 1.
11. M. R. Al-Mamun, S. Kader, M. S. Islam and M. Z. H. Khan, *J. Environ. Chem. Eng.*, 2019, **7**, 103248.
12. K. Winkelmann, T. Noviello and S. Brooks, *J. Chem. Educ.*, 2007, **84**, 709.
13. K. J. Nordell, E. M. Boatman and G. C. Lisensky, *J. Chem. Educ.*, 2005, **82**, 1697.
14. G. Lisensky, R. McFarland-Porter, W. Paquin and K. Liu, *J. Chem. Educ.*, 2020, **97**, 806.
15. N. E. Markina, M. V. Pozharov and A. V. Markin, *J. Chem. Educ.*, 2016, **93**, 704.
16. A. V. Markin and N. E. Markina, *J. Chem. Educ.*, 2019, **96**, 1438.
17. A. D. McFarland, C. L. Haynes, C. A. Mirkin, R. P. Van Duyne and H. A. Godwin, *J. Chem. Educ.*, 2004, **81**, 544A.
18. Citrate Synthesis of Gold Nanoparticles, https://chemistry.beloit.edu/edetc/nanolab/gold/index.html, (accessed May 2021).
19. S. L. A. Paluri, M. L. Edwards, N. H. Lam, E. M. Williams, A. Meyerhoefer and I. E. P. Sizemore, *J. Chem. Educ.*, 2015, **92**, 350.
20. K. A. Duncan, C. Johnson, K. McElhinny, S. Ng, K. D. Cadwell, G. M. Z. Petersen, A. Johnson, D. Horoszewski, K. Gentry, G. Lisensky and W. C. Crone, *J. Chem. Educ.*, 2010, **87**, 1031.
21. J. Soukupova, L. Kvitek, M. Kratochvilova, A. Panacek, R. Prucek and R. Zboril, *J. Chem. Educ.*, 2010, **87**, 1094.
22. S. D. Solomon, M. Bahadory, A. V. Jeyarajasingam, S. A. Rutkowsky, C. Boritz and L. Mulfinger, *J. Chem. Educ.*, 2007, **84**, 322.
23. Basics of Green Chemistry, https://www.epa.gov/greenchemistry/basics-green-chemistry#, (accessed May 2021).
24. Z. Dong, D. Richardson, C. Pelham and M. R. Islam, *Chem. Educator*, 2008, **13**, 240.
25. J. Cooke, D. Hebert and J. A. Kelly, *J. Chem. Educ.*, 2015, **92**, 345.
26. G. Paniconi, S.-J. Chung, S.-C. Liou, S. S. Kim, L. Kim, S. Kang and D. Kim, *Chem. Educ.*, 2016, **21**, 73.
27. A. Nooralhuda, R. A. Alkhalaf, M. C. Neick, P. J. Cleaver and S. Lundsford, *Chem. Educ.*, 2018, **23**, 28.
28. A. Richardson, A. Janiec, B. C. Chan and R. D. Crouch, *Chem. Educ.*, 2006, **11**, 331.
29. R. Vazquez-Muñoz and N. Takeuchi, *J. Nano Educ.*, 2016, **8**, 114.

30. K. M. Metz, S. E. Sanders, A. K. Miller and K. R. French, *J. Chem. Educ.*, 2014, **91**, 264.
31. K. Winkelmann, L. Bernas, B. Swiger and S. Brown, *J. Chem. Educ.*, 2017, **94**, 751.
32. Titanium Dioxide Raspberry Solar Cell, https://chemistry.beloit.edu/edetc/nanolab/TiO2/index.html, (accessed May 2021).
33. Y. R. Smight, E. Crone and V. Subramanian, *J. Chem. Educ.*, 2013, **90**, 1358.
34. S.-I. Chien, C. Su, C.-C. Chou and W.-R. Li, *J. Chem. Educ.*, 2018, **95**, 1167.
35. N. C. Nicolaidis, P. V. Hollott, B. Stanwell, I. A. Gill, J. E. Bull, S. Bentsen, J. Iredale, T. M. Pappenfus, P. C. Dastoor, K. Feron, M. J. Griffith and N. P. Holmes, *J. Chem. Educ.*, 2020, **97**, 3751.
36. P. Berger, N. B. Adelman, K. J. Beekman, D. J. Campbell, A. B Ellis and G. C. Lisensky, *J. Chem. Educ.*, 1999, **76**, 943.
37. A.-L. Dalverny, G. Leyral, F. Rouessac, L. Bernaud and J.-S. Filhol, *J. Chem. Educ.*, 2018, **95**, 121.
38. P. Y. Furlan and M. E. Melcer, *J. Chem. Educ.*, 2014, **91**, 1966.
39. D. VanDorn, M. T. Ravalli, M. M. Small, B. Hillery and S. Andreescu, *J. Chem. Educ.*, 2011, **88**, 1119.
40. D. G. da Silva, F. M. De Melo, A. T. Silveira Jr, B. C. da Cruz, C. C. P. Prado, L. C. P. de Vasconcelos, V. A. S. Lucas and H. E. Toma, *J. Chem. Educ.*, 2016, **93**, 1929.
41. Sachiko Kodama, http://www.kodama.hc.uec.ac.jp/index-e.html, (accessed May 2021).
42. Ferrofluid Tests - Magnetic Liquid, https://www.youtube.com/watch?v=kL8R8SfuXp8, (accessed May 2021).
43. Watch this ferrofluid speaker pulse to the music https://www.cnet.com/news/watch-this-ferrofluid-speaker-pulse-to-the-music/ (accessed May 2021).
44. D. J. Campbell, M. S. Baliss, J. J. Hinman, J. W. Ziegenhorn, M. J. Andrews and K. J. Stevenson, *J. Chem. Educ.*, 2013, **90**, 629.
45. H. Margel, B.-S. Eylon and Z. Scherz, *J. Chem. Educ.*, 2004, **81**, 558.
46. E. W. Ong, A. Razdan, A. A. Garcia, V. B. Pizziconi, B. L. Ramakrishna and W. S. Glaunsinger, *J. Chem. Educ.*, 2000, **77**, 1114.
47. Z. Hajkova, F. Fejar and P. Smejkal, *J. Chem. Educ.*, 2013, **90**, 361.
48. V. Goss, S. Brandt and M. Lieberman, *J. Chem. Educ.*, 2013, **90**, 358.
49. D. Campbell, E. Freidinger, M. Querns, S. Swanson, A. Ellis, T. Kuech, A. Payne, B. Socie, S. M. Condren, G. Lisensky, R. Rassmussen, T. Hollis and R. Villarreal, *Exploring the Nanoworld with LEGO Bricks, Board of Trustees of Bradley University*, Peoria, IL, 2011, pp. 2.1–2.2.
50. D. J. Campbell, J. A. Olson, C. E. Calderon, P. W. Doolan, E. A. Mengelt, A. B. Ellis and G. C. Lisensky, *J. Chem. Educ.*, 1999, **76**, 1205.
51. D. J. Campbell, E. R. Freidinger, J. M. Hastings and M. K. Querns, *J. Chem. Educ.*, 2002, **79**, 201.
52. A. Mulchandani, A. J. Atkinson, S. Garcia-Segura and P. Westerhoff, *J. Chem. Educ.*, 2019, **96**, 708.

CHAPTER 6

Nanochemistry as a Relevant Concept in Teaching Chemistry

6.1 Introduction

According to Gilbert, an expert in science education, to teach science one must deal with several issues:[1]

- Over loaded content – generally required from the teaching curriculum
- Isolated facts – approaches are taught without student understanding on how to draw connections between different facts; this may lead to low class participation and forgetting memorized, isolated facts afterwards.
- Lack of transfer – students can solve problems presented to them in ways they have learned and encountered previously. These same students may fail to solve problems using the previously learned concepts when presented to them in different ways, due to the lack of transfer of what is learned for daily life. Therefore, many students choose not to pursue chemistry studies because they do not see the relevance to what they have learned.[2,3]

One way to deal with these challenges is to create context through teaching techniques that address the social and cultural environment where the student exists.[1] Teaching in context is characterized by employing social, technical, or scientific frameworks to develop an understanding of chemistry. Chemistry then becomes more relevant to the daily lives of the students.[4] Gilbert examined theories based on context-based learning and concluded that the role of teaching in context is to enable students to be provided with meaning and experience. Through meaningful learning

Advances in Chemistry Education Series No. 9
Nanochemistry for Chemistry Educators
By Riam Abu Much, Kurt Winkelmann and Muhamad Hugerat

Published by the Royal Society of Chemistry, www.rsc.org

experiences, the student connects the concepts relevant to aspects of their lives.

Nanoscience and its respective applications are good candidates for teaching in the context of science education because they have clear connections to student daily life, they have industrial aspects and applications, and all represent a rich environment involving research, society, and industry.[4] Nanotechnology education programs have evolved in many places in the world.[5] Studies have shown that the NANOLEAP program, which combines basic concepts in science (such as intermolecular forces) with concepts in nanoscience, enhances student understanding of basic concepts in nanotechnology. This program also improves student understanding of basic scientific concepts and supports research-based learning. Context-based science education with nanotechnological applications may assist a student in developing a better understanding of basic scientific concepts.[6] Within the NANOLEAP program, two teaching modules were developed in which nanoscience concepts could fit into high school physics and chemistry classes in a manner that supports students in learning basic science concepts. One teaching module combines several activities in which the students explore, make observations and interpretations of how the gecko's foot interacts with surfaces. The students concluded that the gecko's foot is composed of arrays of tiny hairs with spatula-shaped tips, each spatula is about 100 nanometres thick. Applying this module, the students learn about static force, intermolecular forces, and how nanoscale components increase the effects of these forces. In the second teaching module, the students explore how properties of materials at the nanoscale differ from those of macroscale samples of the same material. The findings of the NANOLEAP program show that the application of the teaching modules increased understanding of basic science concepts. Other research supports this type of studying as effective, useful, and relevant to daily life Emphasizing the forefront of science and applying a pedagogical technique where the student is at the center can cause an increase in interest and enhanced motivation among students.[7] Elster indicates that context teaching of biology makes biological concepts more interesting, and the students become more motivated.[8]

According to the studies presented, applications of nanotechnology represent scientific concepts that can make science more relevant, interesting, and meaningful to students and their daily lives. Context teaching by applying nanoscientific aspects and applications provide students with an opportunity to learn how modern science works and can encourage students to think about a career in science.[5]

Roco (2002) described the essence of nanotechnology in the following quote:[9]

> "The essence of nanotechnology is the ability to work at the atomic, molecular and supramolecular levels, in the length scale of about 1 to 100 nm range, in order to create, manipulate and use materials, devices

and systems with fundamentally new properties and functions because of their small structures. It includes understanding of phenomena and processes at the nanoscale, and integration of nanostructures along larger scales."

The uniqueness of nanoscience and nanotechnology is that it is a multidisciplinary scientific field which includes concepts in chemistry, biology, physics, engineering, and computer science. These concepts greatly influence various applications of materials, systems, and devices at the nanoscale, as illustrated in Figure 6.1.[10]

The applications of nanotechnology can be considered a technological and medical revolution that serves humanity in various fields including energy, food, environment, medicine, and others. As described in Chapter 1, properties of materials at the nanoscale change dramatically when compared to their larger sized counterparts. This is attributed to two main effects. The first is their high surface to volume ratio and the second is the quantum confinement effect that influences their optical properties.[11–22] These unique properties of nanomaterials lead to many applications that are regularly used in our daily life.

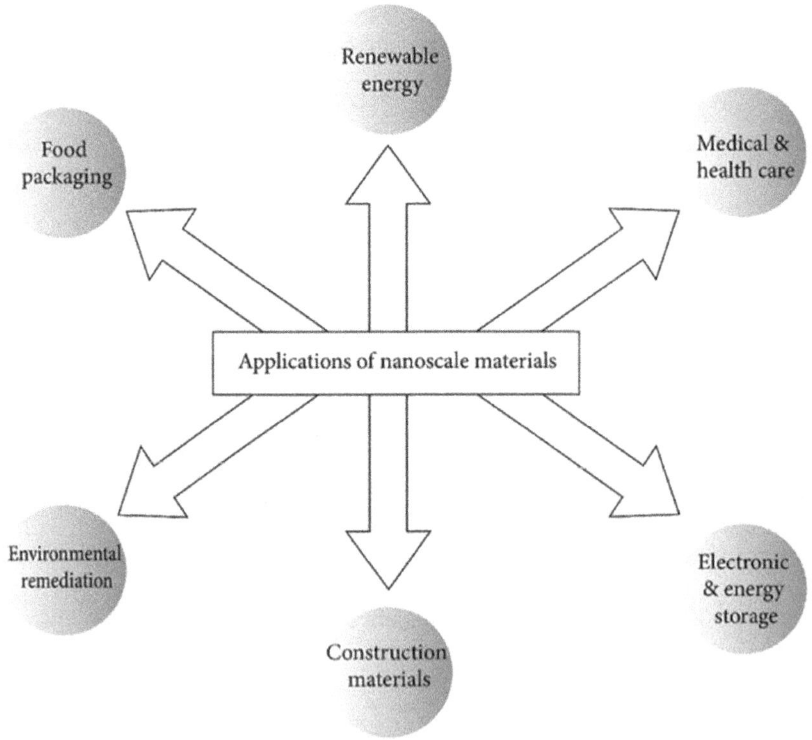

Figure 6.1 Applications of nanomaterials in various fields.

In their research, Blonder and Sakhnini identified seven essential concepts that must be taught in nanoscience lessons (more details are presented in Chapter 4).[23] These concepts are:

1. Size dependent properties
2. Innovation and applications of nanotechnology
3. Size and scale
4. Characterization methods
5. Functionality
6. Classification of materials
7. Approaches of nanomaterials

In later research, they continued to focus on the second essential concept, innovations and applications of nanotechnology that affect our modern daily life, mimic nature, and how they are adapted for several applications.[24] According to Blonder and Sakhnini's findings, they recommend five nanotechnological applications and these are described below.

1. Nanomedicine – the applications of nanoscale structures and materials in medicine are divided into two categories: for medical treatment and diagnosis. Diagnostic applications are based on the change of nanomaterial properties upon binding to the targeted disease site. Medical treatments use nanostructured materials for drug delivery purposes. For example, liposomes and micellar nanostructures act as drug vehicles.
2. Nanoelectronics – nanostructured materials are used in manufacturing electronic devices. For example, carbon nanotubes are used in electronic devices to transfer electrons from one side of the device to the other. This subset of nanotechnological applications is highly connected to daily student life because most students in industrialized areas regularly utilize computers, cellular phones, and LED-based appliances.
3. Photovoltaic cells – students can connect to this topic by learning about the concept of surface area and its importance in increasing the material reactivity and structures at the nanoscale. For example, to increase the efficiency of solar cells, nanomaterials such as titanium oxide are used because of this material's large surface area.
4. Nanobots – this technology relates to the design of nanoscale robots made of nanomaterials and nanostructures that are considered a challenge for future developments.
5. Self-cleaning – this nanotechnological application is closely related to the topic of green environment that is highly relevant to modern daily life. The useful properties of nanoscale materials and their high surface area is exploited for use in different kinds of self-cleaning products. For example, one can fabricate superhydrophobic surfaces by using films made of titanium oxide nanoparticles to produce a photocatalytic effect.

This can be done by allowing sunlight to create oxygen species on the surface of titanium oxide films which can attack organic contaminants. This prevents pollutants from sticking to the surface.

6.2 Relevant Aspects in Teaching Nanotechnology

6.2.1 Nanoscience and Nanotechnology Teaching Module

A teaching module was designed by Blonder for ninth grade students to teach nanochemistry.[25] The teaching module consists of the application of several teaching methods that aim to engage students and make the learned concepts more interesting and motivational. The module includes game-based learning, visualization and multimedia, learning with movies, learning with simulations, and learning with models.[26–33] Blonder's teaching module aims to teach two basic nanotechnology concepts. It incorporates different teaching methods where students can learn about (i) size and scale and (ii) surface area to volume ratio.

The proposed teaching module can be useful for school teachers in order to expose students to basic aspects of nanoscience and nanotechnology. The various teaching methods in this module keep students engaged. The final stage of the module consists of teaching the relevance of nanostructured materials to our daily life. In Blonder's proposed module, the following topics are taught:

- Introduction to nanotechnology
- Introduction to the nanoscale
- The significance of nanotechnology
- Different nanomaterial types and shapes
- Nanotubes, nanoparticles, and their applications
- Surface area to volume ratio
- Applications of nanotechnology and its relevance to daily life.

In accordance with the proposed module designed by Blonder, the following activities are highly recommended to address the subject of nanoscience and nanotechnology to middle and high school students.

Activity 1: Nano scraps (game-based learning) – in this activity students are asked to cut the smallest possible piece of paper and write on it, according to their own knowledge, the name of an object that is in the nanoscale. The aim of this activity is to investigate the preliminary knowledge base students have regarding nanoscience and nanotechnology.[26]

Activity 2: Size and scale – this activity is based on the strategy of learning through visualization. Student are exposed to the relative size of different, familiar objects. This activity allows students to investigate how small the nanoscale is and to correct misconceptions about nanometre size and scale. Figure 6.2 presents how this activity can be executed by showing how students can compare different objects from differing size scales.

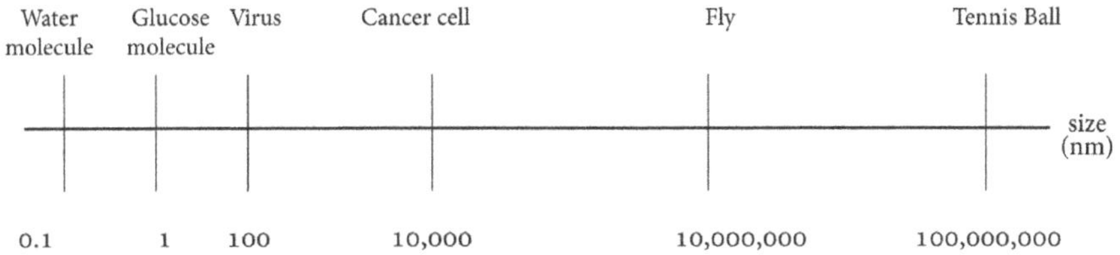

Figure 6.2 A comparison of different size of objects.

Activity 3: Cutting down to nanoscale – this activity employs a game-based learning approach.[26] Students are asked to try and cut a piece of paper to smaller pieces using scissors. First, they are asked to cut the paper in half and then to continue to cut the paper into the smallest piece they can achieve. They will not succeed in cutting the paper to reach the nanometre scale. However, in this activity students experience visually how the paper exponentially decreases in size. To make this game-based activity more interesting and engaging, students can compete for who can cut the smallest piece of paper. This activity demonstrates to students how nanoscale objects are much smaller than the smallest visible objects. It is highly recommended to give examples of small, invisible objects that can have large impacts, such as the coronavirus.

Activity 4: What is nanotechnology? – this activity is based on the strategy of learning with the aid of videos and movies.[30,31] Students watch a video where two children explain what nanotechnology is.[34] In the video, a human hair is enlarged to model how small nanoscale objects are in size. This activity exposes students to the nanoscale and nanotechnology in a friendly, relatable way in order to explain the amazing world of nanotechnology.

Activity 5: Significance of nanotechnology – this activity is also based on the strategy of learning with the aid of videos and movies.[30,31] This activity employs the use of short movies to demonstrate to students the variety of uses nanotechnological applications have in industry, technology, and their importance to daily life. An appropriate movie selection will actively engage students and introduce them to unusual or lesser-known nanotechnology applications.[35,36]

Activity 6: Learning with dice – this activity uses models to demonstrate the main effects that surface area to volume ratios have on materials and structures at the nanoscale.[33] The students use dice, cubes, and Hungarian cubes (known as the Rubik's Cube) as tangible models to understand the concept of surface area per volume.

Activity 7: Material polystyrene model – this activity is based on learning with models. The students build a cube from polystyrene balls and each ball represents an atom. From this, students can realize the importance that surface area plays in different physical and chemical properties of materials. They will obtain a better understanding of how atom arrangements on the surface differ when they are surrounded by fewer atoms and exposed to the environment. They are surrounded by fewer atoms in comparison to atoms and molecules inside the material.

Activity 8: Models and phenomena from daily life – this activity connects phenomena and models from our daily life to concepts of surface area per volume ratios. For example, the wrinkles in elephant skin increases its surface area, and the wrinkles aid in spreading the heat over a larger surface area to keep the animal cooler. Some heat radiators are designed similarly, where the radiators which are divided into rectangles enhance the surface area, and then the heat dissipates more effectively.

Activity 9: Building different types of nanostructured materials – the activity is based on models and game-based learning strategies. Students use models to build fullerene and carbon nanotubes to assemble different types and shapes of nanostructured materials. This activity can also be accompanied with presenting different three-dimensional images of nanomaterials.

Activity 10: Nanotechnology applications – in this activity students are exposed to different kinds of nanotechnology applications that are familiar to them, for example, technological gadgets like iPhones. Electrodes of iPhones are coated with nanoparticles in order to increase the available electrode surface area. As a result, this increase in surface area enhances the power available to the battery and lessens the time required for battery recharge.

The implementation of the above activities was evaluated by examining students' perceptions and understanding concepts that are relating to nanoscale. The students expressed positively regarding the different activities. The results indicated that students had poor pre-knowledge about the nanoscale, whereas, after the application of the different activities, the students could give examples regarding optional objects that are at the nanoscale.

6.2.2 The Self-cleaning of Lotus Leaves: Superhydrophobic Surfaces

The leaves of lotus plants fascinate many scientists due to their ability to repel dirt and clean themselves. Large amounts of research have been dedicated to elucidating the structure of lotus leaves and the mystery behind their ability to self-clean.[37,38] This unique property was thought to be due to their superhydrophobic surfaces with roughness caused by microfilaments.[39] Recent studies reported that the surface of a lotus leaf consists of micro- and nanoscale hierarchical structures, or branch-like nanostructures on the top of micro-papillae.[40] Figure 6.3 presents an image of a water bubble on a lotus leaf. The shape of the water droplet does not spread across the surface of the leaf, thereby indicating that the lotus leaf possesses hydrophobic and water-repellent properties.

The remarkable ability of lotus leaves to repel dirt has been an inspiration for developing disinfection and self-cleaning technologies. The discovery of rough-surfaced lotus leaves containing branch-like nanostructures played a role in the fabrication of superhydrophobic surfaces based on hydrophobic nanostructures. Yuyang *et al.* fabricated artificial lotus leaf structures by assembling carbon nanotubes onto cotton surfaces. Inspired by the natural lotus

Figure 6.3 An image of water droplets on lotus leaf.

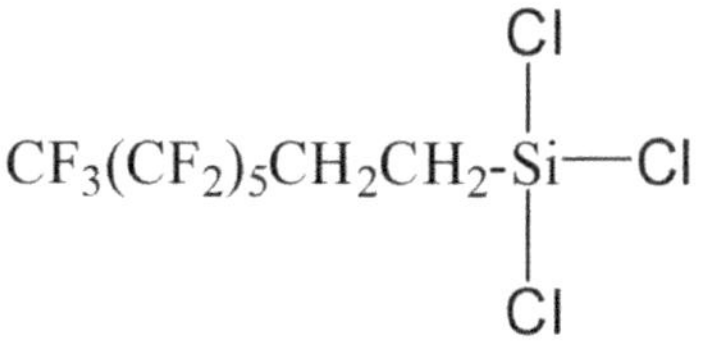

Figure 6.4 Chemical structure of trichloro(1*H*,1*H*,2*H*,2*H*-perfluorooctyl).

leaf structure, the authors attempted to convert water absorbent cotton fabrics to display superhydrophobic surfaces.[41]

Recently, Muharjan *et al.* applied a self-cleaning, hydrophobic nanocoating onto glass substrates.[42] They began the process by roughing flat glass surfaces with micro and nanoscale particles of diamonds to increase surface area. This was an idea inspired by the natural surfaces of lotus leaves. This step was followed by the deposition of the hydrophobic trichloro(1*H*,1*H*,2*H*,2*H*-perfluorooctyl) silane – its chemical structure is presented in Figure 6.4.

Another study created lotus-like superhydrophobic copper substrates without chemical modification. This process was based on roughing copper plates *via* the electroplating method.[43,44]

The application of superhydrophobic surfaces are widely used in modern daily life as well as for uses in industrial processes such as textiles, cleaning, and corrosion protection.[45–47] The natural structure of lotus leaves with its nanoscale effects became an inspiration for forming superhydrophobic surfaces for several purposes.

6.2.2.1 Converting a Copper Plate Surface to a Hydrophobic One – Lab Experiment

Lotus plant leaves and their superhydrophobic, rough surfaces based on branch-like nanostructures can be introduced as a laboratory activity. Here students convert hydrophilic copper plates to become hydrophobic in order to imitate lotus leaf structures.[44] The experiment consists of two steps. In the first step copper plates are roughed and the second step involves coating the surface of the copper plate with hydrophobic material. During this activity, students can compare the wettability of the copper plate in three different situations: before roughing the copper plate; after roughing; and after coating with hydrophobic material. The experimental procedure is described below.

Experimental procedure

Materials: Copper plates (2×2 cm^2), aqueous solution of 2.0 M NaOH, aqueous solution of 1.0 M $K_2S_2O_8$, 5 mM ethanol solution of dodecanoic acid, and acetone.

Step 1: Roughing the copper plate surface – immerse copper plates into an aqueous solution of 2.0 M NaOH and 0.1 M $K_2S_2O_8$ at room temperature for 20 minutes. Take the copper plates from the solution and rinse them thoroughly with distilled water. Air dry. This process roughens the copper plate surface by creating copper hydroxide nanowires with interconnected architectures that cover the copper surface.

Step 2: Coating the surface of the copper plate with hydrophobic material – immerse the copper plate from step one into 5 mM ethanol solution of dodecanoic acid for another 20 minutes. This process creates self-assembled monolayer of copper carboxylate. Rinse the copper plates with acetone. Air dry.

6.3 Coating with Nanomaterials: The Magic Sand Activity

When teaching chemistry, materials can be categorized according to their hydrophobic properties as being either hydrophilic or hydrophobic. Hydrophilic materials are defined as materials that attract water. Hydrophobic materials repel water molecules. The differing chemical structures are what determine if a material is hydrophilic or hydrophibic.[48]

The magic sand activity can demonstrate examples of hydrophilic and hydrophobic materials along with how nanoscale materials affect surface properties. The standard sand types are made of hydrophilic materials that are strongly attracted to water molecules. This phenomenon is clearly observed at the beach when sand becomes wet with sea water and it turns into mud. Scientists can use nanotechnological means to coat natural sand grains with silicon-based compounds. This silicon-based compound can possess a hydrophilic end on one side and a nanoscale-sized hydrophobic end on the other. The hydrophilic end attracts the sand grains. The hydrophobic layer over the other end of the sand grain repels water. This layer would convert standard sand from strongly attracting water molecules to strongly repelling them.[49] The activity can be designed for high school students when teaching the traditional chemical subject of intermolecular

Figure 6.5 An image of dry magic sand in water.

forces or for middle school students when teaching about hydrophilic and hydrophobic materials.

Magic sand can be used as a simple experiment to demonstrate how "unseen" nanoscale layers can dramatically affect the chemical behavior of materials. By coating the sand with specific nanomaterials, students can compare between regular sand and magic sand and how the sand interacts with water. The magic sand activity is very simple, with the materials needed being only natural sand, magic sand, and water. Water is poured into two beakers or cups: one cup contains natural sand and the other cup contains a certain quantity of magic sand. Students can then witness each respective cup of sand's interaction with water. The students will observe that magic sand remains dry while natural sand becomes wet, as illustrated in Figure 6.5.

6.4 Teaching Nanotechnology Concepts to Elementary Students – Surface Area

When teaching about nanoscience and nanotechnology, it is very important to address two basic phenomena that are responsible for the principal changes of the physical and chemical properties of materials when they are at the nanoscale. As described in Chapter 1, a large surface area to volume ratio is a principal feature of nanoscale sized materials.

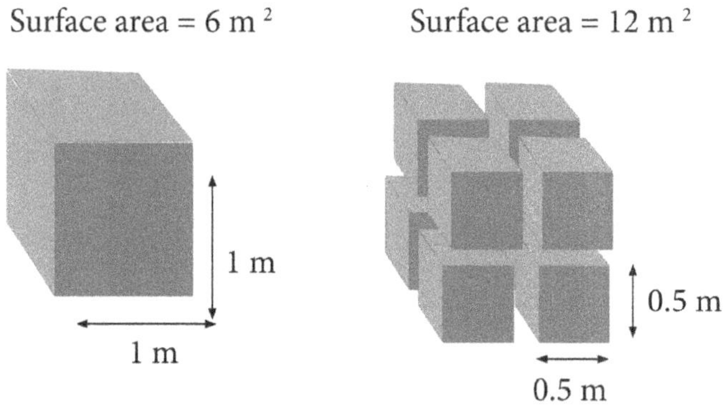

Figure 6.6 Calculation of surface area of cube and the total surface area after divided it to eight cubes.

Teaching nanotechnology needs to be based on understanding its basic aspects and its importance in daily life.[50] In research performed by Nandiyanto, he and his team performed a very interesting experiment that could be considered as a simple starting point to expose students to basic aspects of nanotechnology.[51] The suggested experimental demonstration is based on the correlation between the size of solute particles and their dissolution into water, a phenomenon that is highly relevant to modern daily life. This experiment puts the surface area aspect of nanoscale materials into context. For elementary students, dissolution is explained plainly, without mentioning more complicated chemistry concepts such as ion formation, concentration of solution, and chemical bonds, among others. Nandiyanto and his team relied on the definition that dissolution could be explained by homogenously distributing solute particles in the solvent phase. Solute particle size significantly affects the dissolution ability of the particles, as smaller particles dissolve more easily than larger ones. In order to explain this in a simple and clear way, a cube is presented, and a comparison is shown between the obtained surface area and the total surface area after cutting it into several smaller cubes, as shown in Figure 6.6. A similar and more extended activities can be found in activity number 6 above in this chapter and in Chapter 5, Activities for Teaching Nanochemistry.

Two kinds of solutes were used, sugar and table salt. Both solute types were of differing sizes as shown in Figure 6.7.

As described above, the size of solute particles affects its dissolution ability, the hypothesis is that cutting the solute particle to smaller ones, increases the dissolution rate. This activity can be useful for learning about the effects of surface area on the reactivity of materials. Surface area size is one of the basic factors that affect the properties of materials at the nanoscale. It is highly recommended to expand the scope of this experiment by relating it to other phenomena from modern daily life that are influenced by the relationship between surface area and material properties. For example, students could

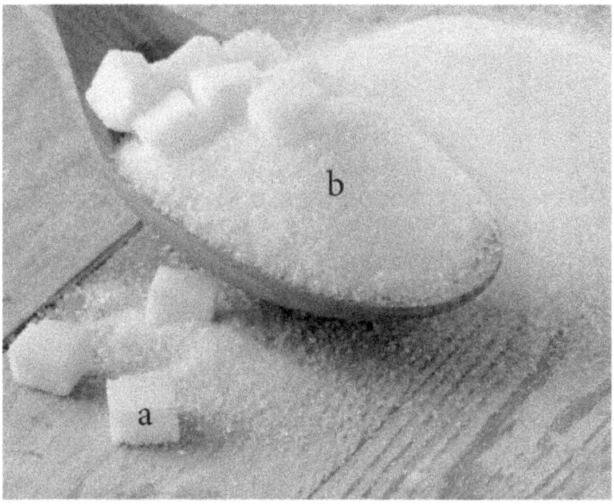

Figure 6.7 An image of different sizes of sugar particles, a sugar cube, b, sugar powder.

Figure 6.8 Burning of a piece iron nail and iron wool.

compare between burning a piece of iron *versus* a piece of iron wool, as illustrated in Figure 6.8. Another example could include melting ice cubes *versus* melting ice cubes of the same volume after grinding them into very small pieces. Evaporation of a known volume of water *versus* the evaporation of the same volume after dividing it into small quantities.

6.5 From Optical Properties of Gold Nanoparticles to Colorimetric Sensors

Metal nanoparticles are characterized by the unique colors they exhibit that are different from their known color when sized in the macroscale. As described in Chapter 1, colloidal solutions of metal nanoparticles are characterized by localized surface plasmon resonance (LSPR). Briefly, the small size of metal nanoparticles is consistent with the wavelength of visible light. Therefore, an interaction between the light and the metal nanoparticles occurs which produces an oscillation of the metal nanoparticle free electrons. This phenomenon results in the absorption and emission of UV-visible light that produces a spectrum of visible colors in metal nanoparticle solutions.

As described in Chapter 1, metallic nanoparticles are known for their unusual and beautifully bright colours. Thus, metallic nanoparticles are used for developing sensors that are based on visualizing color changes.[52–55] LSPR is strongly affected by both the size and shape of metal nanoparticles. As a result, different sizes of metal nanoparticles could exhibit differing colors.

As described in Chapter 1, Section 1.6.2, the effect of the larger surfaces area per volume of materials at the nanoscale causes van der Waals attraction forces to be great influencers, which may lead to nanomaterial aggregation. In order to prevent this from happening, different stabilization methods have been developed. Electrostatic stabilization is one such method that is based on bestowing an electric charge to the nanoparticles in order to overcome the attraction forces between the nanoparticles.[56]

McFarland and his colleagues proposed an undergraduate practical lab that introduces students to the unique optical properties of gold nanoparticles.[57] In this activity, students prepare aqueous colloidal solutions of gold nanoparticles *via* applying the reduction method. The reducing agent acts as both a reducer and stabilizer by adsorbing onto gold nanoparticle surfaces and thereby creating a negative electrical charge. Next, students explore the applicability of using gold nanoparticles to act as an electrolyte sensor. This is done by adding electrolyte solutions which induce gold nanoparticle aggregation by enhancing nanoparticle electrical attractive forces (see Section 1.6.2.1.1 in Chapter 1). This process produces color changes in the colloidal solution that is attributed to the nanoparticle size changes induced by the added reagents. The activity can be applied when teaching topics of reduction oxidation reactions, electrolytes, and complementary colors.

6.5.1 Developing Electrolyte Sensors from Gold Nanoparticles

This lab activity can be to prepare an aqueous colloidal solution which acts as a sensor for the detection of different solutions. The preparation method is based on the reduction of gold ions by trisodium citrate ($Na_3C_6H_5O_{7(aq)}$).[58] The preparation procedure is described schematically in Figure 6.9. In this figure, an aqueous solution of $HAuCl_4$ is heated to boiling temperatures while continually stirred. After the solution begins to boil a trisodium citrate solution is added. Stirring and heating are continued until the solution turns red in color, which indicates the presence of gold nanoparticles.[57]

In Figure 6.9, trisodium citrate reduces gold ions to gold atoms that accumulate to produce gold nanoparticles. In an aqueous solution, trisodium citrate has a negative charge. The excess trisodium citrate in the solution adsorbs to the gold nanoparticle surfaces, conferring upon them an overall negative electrical charge.

Gold nanoparticle solutions are characterized by their differing color displays that depend on the size and shape of the gold nanoparticles. Differently sized nanoparticles determine how they interact with light and this is visually reflected by the produced color. In fact, red-colored solutions of gold nanoparticles turn to purple when the nanoparticle sizes increase. The unique feature of size-dependent color makes negatively charged gold nanoparticles a form of selective chemical sensor.

In one example, an activity can be performed to differentiate between a table salt solution and a sugar solution. The two solutions appear to look the same, but chemically, table salt (NaCl) is an ionic compound and sugar is molecular. Therefore, their dissolution modalities in water are different. No solute ions are available in the sugar solution, whereas the solution of table salt contains positive and negative ions. Based on these facts, the addition of sugar solution to the solution of red colored colloids made of gold nanoparticles will not lead to any color change. In comparison, adding the salt solution to the gold nanoparticle solution will lead to the gradual transition of its colors to turn from red to purple. Precipitation of the aggregated gold nanoparticles may also happen at this time. This color change occurs due to the presence of positively charged sodium ions that reduce the intensity of the negative charges of the gold nanoparticles and the attraction between them. This leads to increasing the gold nanoparticle sizes and this is what causes the color change.

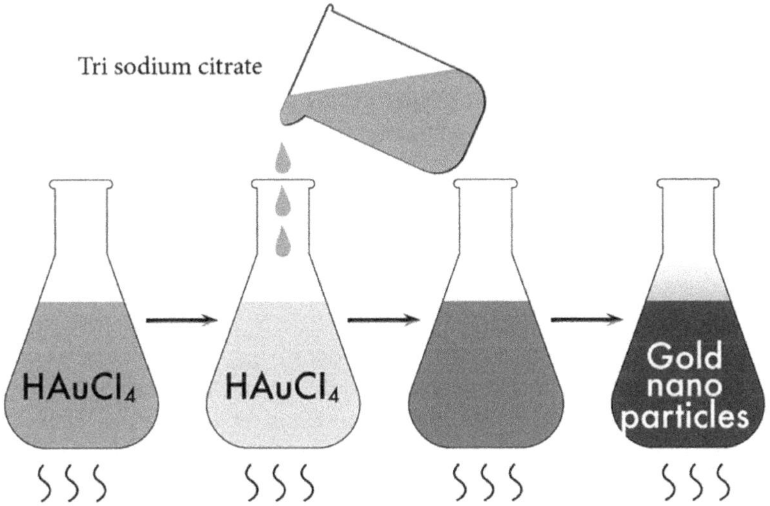

Figure 6.9 Schematic description of the preparation process of gold nanoparticles by reduction with trisodium citrate.

6.6 Preparation of Magnetic Liquids at Low Temperatures

Conventional methods to create liquified magnetic material require extreme conditions of high temperatures to melt a highly magnetic solid-like iron. There are two principle challenges that conventional methods present in order to accomplish this goal. The first challenge is that in order to keep magnetic material in a liquid state, a high temperature must be applied at all times. This makes it difficult to use for different purposes. The second challenge is that solid magnetic materials can lose degrees of magnetism upon melting. This happens because the material magnetic properties are strongly affected by temperature. Each magnetic material has a specific Curie temperature (T_C), a property that is attributed to magnetic materials where thermal energy overwhelms the tendency of electrons to align in a magnetic domain. In most cases, T_C is lower than the melting point.[59,60]

The development of nanoscience and nanotechnology has contributed to many positive changes that are reflected in our daily life. The production of different nanoscale materials has had a great impact in many fields. As mentioned in Chapter 1, ferrofluids are differing colloidal solutions made of magnetic nanoparticles. Interestingly, the production of ferrofluids overcomes the obstacles mentioned above.

Ferrofluids are liquid magnetic solutions that are stored at room temperature and can be controlled by an external magnetic field. Ferrofluids have a wide range of applications both for technological and medical purposes. The American National Aeronautics and Space Administration (NASA) developed the first ferrofluid in 1960 and used it for rotating shaft seals in satellites. Nowadays, ferrofluids perform a significant role in a variety of machines ranging from centrifuges to computer hard drives.[61] Controlling ferrofluids using external magnetic fields makes them useful materials for various medical applications. For example, an actuator made with ferrofluids as a part of its design is used for an implantable artificial heart. The actuator can be controlled by simply applying an external magnetic field.[59]

Aqueous magnetic fluids have also been successfully used as drug vehicles. Drugs are attached to the surface of magnetic nanoparticles and, *via* applying a magnetic field, the drug can be directed to a specific desired site in the body.[62] Makhluf *et al.* successfully loaded magnetite nanoparticles into sperm cells. After this process the sperm cells kept their mobility and ability to fertilize an egg.[63,64] This process made sperm cells acquire a magnetic property that enabled the management of their movement by applying an external magnetic field. In principle, magnetic nanoparticles of ferrofluids can be bonded to various structures and biomaterials in order to control their motion and activity for different purposes.

In the United States, one can identify an authentic dollar bill by its attraction to a strong magnet. Ferrofluids are used as magnetic ink in the ink jet printers used to manufacture the bills and therefore magnets can verify their authenticity.[65]

Berger and his colleagues proposed a practical activity that describes the preparation and characterization of ferrofluids.[65] The activity can be performed with high school students or undergraduate students. In this activity, magnetite Fe_3O_4 nanoparticles are synthesized and then stably dispersed in an aqueous solution. A magnetite crystallized structure is composed of both Fe(II) and Fe(III) ions. The preparation of magnetite is based on the reaction between both iron ions in an aqueous ammonia solution, as described in the following reaction:

$$2FeCl_3 + FeCl_2 + 8NH_3 + 4H_2O \rightarrow Fe_3O_4 + 8NH_4Cl \tag{6.1}$$

The process of preparing colloidal solutions involves two important factors. The first factor involves finding the appropriate preparation method for the nanoscale material. The second factor is attributed to the process of stabilization. This means it is an important to find a stabilizing agent which can provide enough repulsive forces between the nanoparticles so that the nanoparticles are not attracted to one another.

The preparation of a stable dispersion of magnetite (Fe_3O_4) nanoparticles in an aqueous solution is considered a larger challenge because of the magnetite's innate magnetic properties. As described in Chapter 1, the large surface area of materials at the nanoscale makes van der Waals attraction forces more powerful. Therefore, a stabilization mechanism should be applied in order to overcome these stronger attraction forces. In addition to van der Waals forces, strong magnetic attraction forces exist that makes the stabilization process of magnetite more difficult.[66–69] The suggested preparation method of Berger and his colleagues can be summarized in the following schematic diagram in Figure 6.10.

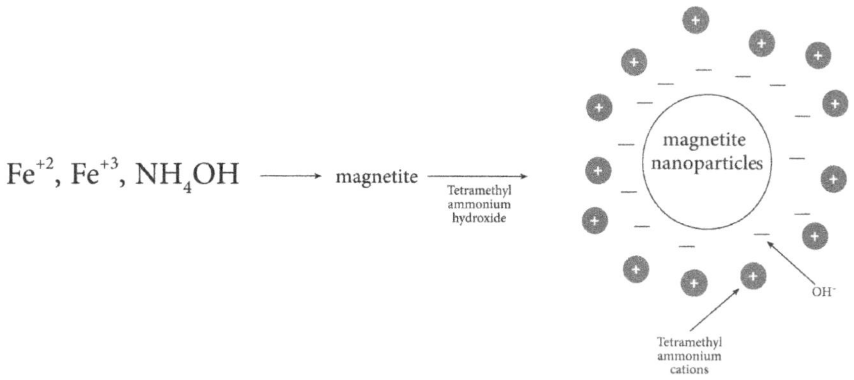

Figure 6.10 Schematic description of the preparation process of magnetite nanoparticles.

6.6.1 Preparation of Ferrofluid-lab Experiment

Materials types and their respective quantities needed to prepare a ferrofluid experiment are described in Table 6.1.[65]

1. In an Erlenmeyer flask, combine 1 mL of $FeCl_2$ and 4 mL of $FeCl_3$ solutions and stir vigorously using a magnetic stirrer.
2. Slowly drip 50 mL of NH_3 solution into the mixed solution of $FeCl_2$ and $FeCl_3$.
3. After five to ten minutes, a black precipitate will settle down into the flask. This black precipitate indicates the formation of aggregated magnetite nanoparticles.
4. The next step involves the preparation of stable magnetite nanoparticles. Here the nanoparticles are stabilized against aggregation, and this is when ferrofluid is produced.
5. Separate out the magnetite precipitate and the solution above, preferably by centrifugation.
6. Disperse the magnetite precipitate into a solution of tetramethylammonium hydroxide $(CH_3)_4NOH$ *via* stirring.
7. In order to remove excess ammonia from the solution, stir the solution using a magnetic stirrer and place under a vacuum aspirator for 30 minutes.

An aqueous solution of tetramethylammonium hydroxide $(CH_3)_4NOH$ consists of hydroxyl ions (OH^-), and tetramethylammonium ions $N(CH_3)_4{}^+$. The stabilization of magnetite nanoparticles can be achieved in two successive steps. In the first step, hydroxyl ions surround magnetite nanoparticles by adsorbing onto the nanoparticle surfaces. This adsorption occurs because magnetite nanoparticles are composed of both Fe^{3+} and Fe^{2+} ions. In addition, tetramethylammonium cations adsorb to the surface of magnetite nanoparticles *via* attractive forces between positive and negative charges. As illustrated in Figure 6.10, the stabilization process of magnetite nanoparticles results from electric reduction forces that arise from the positively charged tetramethy-lammonium ions.[65]

The importance of ferrofluids and their numerous applications for technology and biomedicine are related to their unique physical property of

Table 6.1 Material types and their respective quantities needed to prepare a ferrofluid.

Materials	Details
1.0 M $FeCl_3$ in 2 M HCl	67.58 g of $FeCl_3 \cdot 6H_2O$ in 250 ml of 2 M $HCl_{(aq)}$
2.0 M $FeCl_2$ in 2 M HCl	39.76 g of $FeCl_2 \cdot 4H_2O$ in 100 ml of 2 M $HCl_{(aq)}$
0.7 M $NH_{3(aq)}$	48 mL of concentrated aqueous solution of NH_3 diluted to 1 L with H_2O
25% aqueous solution of $(CH_3)_4NOH$	

being able to be physically manipulated by an external magnetic field. The preparation of a magnetite ferrofluid could also be followed by a simple visual illustration of the effect of an external magnet (see Figure 5.4, Chapter 5).

6.7 Nanoscale Structure-based Light Emitting Diode (LED)

Blonder designed a teaching module for high school students based on teaching the topic of a light emitting diode (LED), its working principals, and the role LEDs play in the development of nanotechnology.[7] LED is a light source composed of semiconductor materials. Unlike conventional light sources that are based on converting electrical energy into heat and then into light, LED is based on converting electrical energy directly into light. Semiconductor materials are characterized by their significant band gap energy. By applying electrical energy, electrons move from the higher energy conduction band to the lower energy valence band. In this transition across energy bands, they emit wavelengths of light from the ultraviolet, visible, and infrared spectrums. The wavelength of the emitted light is based on the energy difference between the conduction and the valence energy states which leads to different colors of light emitted.

As described in Chapter 1, one type of zero-dimensional nanostructure is quantum dots. Quantum dots are very small nanostructures made of semiconductor materials. They are characterized by the phenomenon of quantum confinement, or a higher band gap energy in comparison to microscale semiconductor materials.[70] In addition, quantum dots that are made of the same material could possess a different band gap energy depending on the dot morphology and size on the nanoscale. Smaller sizes of quantum dots possess greater band gap energies. A unique feature of quantum dots is related to the fact that their band gap energy is consistent with the energy of light that is in the visible region. When applying energy, electrons in the valence band "jump" to a higher energy level. When it returns, they emit an energy with a wavelength in the visible light spectrum. The color of emitted light from this occurrence depends on its energy band gap. According to the quantum confinement effect, smaller sized quantum dots emit shorter wavelengths of light because of their higher energy band gap. It is not uncommon to see blue light produced, while red-colored light has been more frequently seen from the emission of smaller sized quantum dots. As shown in Figure 6.11, quantum dots with intermediate-sized, green-colored light can appear.

Blonder adapted the student-centered pedagogy as a teaching strategy for incorporating nanotechnology concepts when teaching chemistry.[7,71] According to her study, increasing student motivation to study chemistry could be achieved by bridging the gap between modern technological developments that affect our daily life and current high school learning platforms.

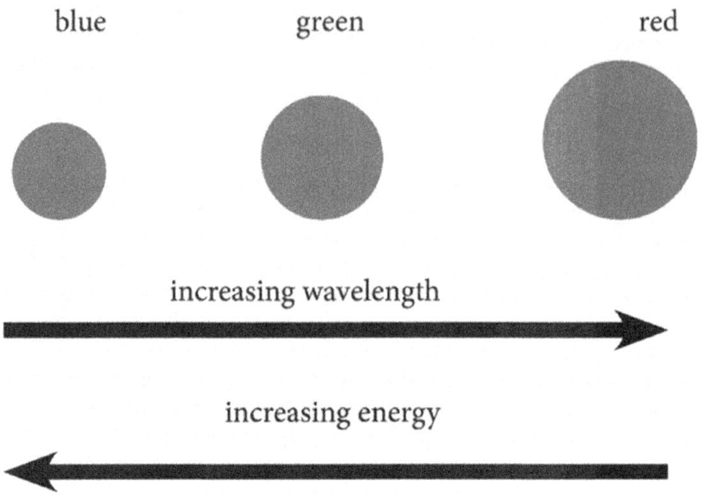

Figure 6.11 Schematic description of different sized of quantum dots that appear to be different colors.

Nanotechnology concepts can be an interesting candidate for teaching about modern life.

6.7.1 Learning Module: LED as a Well-known Nanotechnology Application

This module consists of three different activities which aim to gradually build basic knowledge and highlight features of LED as a modern nanotechnological application.[7]

Activity 1: Diodes as components in an electrical circuit

This activity is composed of two stages. In the first stage, students measure the conductivity of different components and materials and examine the direction of the conductivity. The aim of this activity is to expose the students to different types of materials, insulators, semiconductors, and conductors. In addition, students gain the knowledge that a diode is an electrical circuit component which allows electricity to flow in one direction. By performing these experiments, the students are able to ask questions related to how electrical circuits function. In the next stage of the activity, students are given adapted scientific texts about electrical circuits and their components. For example, they can be given information about silicon as a semiconductor material, its properties, and its behavior.

Activity 2: Color mixing *versus* light mixing

This activity attempts to introduce the topic of light as one type of electromagnetic wave and that light color depends on its energy. During this activity students are asked to mix gouache paints and create different colors.

Then students are asked to try and to create the same colors by mixing different colored light beams. Students discover that mixing different-colored light beams does not always replicate the same colors as the gouache paints can do. This activity can trigger a student's curiosity regarding the concepts of light and color. This is an opportunity to explain that the light's source of color depends on the electronic structure and energy levels of different materials. In the second part of this activity the instructor can perform a demonstration by burning differing metal salts to create different colors in a fire flame. This activity can be followed by another demonstration where students can build an electrical circuit. In this activity, students can compare between a conventional diode that converts electricity to heat and a LED that converts electricity directly into light.

Activity 3: Standard Lamp Efficiency *versus* LED Lamp

In a third activity, students can compare the efficiencies of standard lamp and a LED lamp *via* illuminating an object within a black box. At the end of these three activities, students can receive a concluding assignment that includes scientific text about LEDs.

According to Blonder, this module designed for high school students can expose them to one of the more well-known, advanced applications in nanotechnology – the LED. The role of LEDs in electronic devices is based on the unique properties of nanostructured materials, specifically their large surface area and distinctive optical properties. The study indicated that the application of this activity increased the motivation of students to learn new concepts like LED and concepts in nanotechnology.

References

1. J. K. Gilbert, *Int. J. Sci. Educ.*, 2006, **28**(9), 957.
2. R. Millar and J. Osborne, *Beyond 2000: Science Education for the Future, King's College London*, United Kingdom, 2000.
3. F. Rutherford and A. Ahlgren, *Science for all Americans*, Oxford University Press, New York, 1990.
4. S. Sakhnini and R. Blonder, *Int. J. Sci. Educ.*, 2016, **38**(3), 521.
5. G. Jones, R. Blonder, G. Gardner, V. Albe, M. Falvo and J. Chevrier, *Int. J. Sci. Educ.*, 2013, **35**(9), 1490.
6. A. Greenberg, *ACS Nano*, 2009, **3**(4), 762.
7. R. Blonder and M. Dinur, *J. Nano Educ.*, 2011, **3**(1–2), 51.
8. D. Elster, *J. Biol. Educ.*, 2009, **43**(2), 53.
9. M. C. Roco, *Int. J. Eng. Educ.*, 2002, **18**(5), 488.
10. M. Shariat and F. Ali Akbar, *J. Civil Eng. Urban*, 2016, **6**(4), 84.
11. M. Runowski, *Chemik*, 2014, **68**(9), 766.
12. S. Polizzi, M. Battagliarin, M. Bettinelli, A. Speghini and G. Fagherazzi, *J. Mater. Chem.*, 2002, **12**(3), 742.
13. X. Battle and A. Labarta, *J. Phys. D: Appl. Phys.*, 2002, **35**(6), R15.
14. K. Ostrikov, E. C. Neyts and M. Meyyappan, *Adv. Phys.*, 2013, **62**(2), 113.

15. D. Astruc, F. Lu and J. R. Aranzaes, *Angew. Chem., Int. Ed.*, 2005, **44**(48), 7852.
16. Y. Liang, Y. Li, H. Wang, J. Zhou, J. Wang, T. Regier and H. Dai, *Nat. Mater.*, 2011, **10**(10), 780.
17. Z. Tang, N. Kotov and M. Giersig, *Science*, 2002, **297**(5579), 237.
18. H. Chander, *Mater. Sci. Eng. R Rep.*, 2005, **49**(5), 113.
19. *New and Future Developments in Catalysis, Catalysis by Nanoparticles*, ed. S. Steven, Elsevier, Amsterdam, The Netherlands, 2013, **vol. 9**, pp. 213–244.
20. S. Baskoutas and A. F. Terzis, *J. Appl. Phys.*, 2006, **99**(11), 013708.
21. T. Voddmeyer, D. J. Katsikas, M. Giersig, I. G. Popovic, K. Diesner, A. Chemseddine, A. Eychmuller and H. Weller, *J. Appl. Phys.*, 2001, **90**, 265.
22. V. N. Soloviev, A. Eichhoefer, D. Fenske and U. Banin, *J. Am. Chem. Soc.*, 2000, **122**(11), 2673.
23. S. Sakhnini and R. Blonder, *J. Sci. Educ.*, 2015, **37**(11), 1699.
24. S. Sakhnini and R. Blonder, *Int. J. Sci. Educ.*, 2016, **38**(3), 521.
25. S. Sakhnini and R. Blonder, *Chem. Educ. Res. Pract.*, 2012, **13**(4), 500.
26. S. Barab, M. Thomas, T. Dodge, R. Carteaux and H. Tuzun, *Educ. Technol. Res. Dev.*, 2005, **53**(1), 86.
27. A. Paivio, *Mental Representations: A Dual Coding Approach*, Oxford University Press, New York, 1986.
28. A. Paivio, *Can. J. Psychol.*, 1991, **45**(3), 255.
29. R. E. Mayer, *Multimedia Learning*, Cambridge University Press, New York, 2001.
30. J. V. Ebenezer, *J. Sci. Educ. Technol.*, 2001, **10**(1), 73.
31. R. M. Kelly and L. L. Jones, *J. Sci. Educ. Technol.*, 2007, **16**(5), 413.
32. R. Kozma and J. Russell, *Visualization in Science Education*, ed. J. Gilbert, Springer, Dordrecht, 2005, 7, pp. 121–145.
33. J. K. Gilbert and C. J. Boutler, *Developing Models in Science Education*, Springer, Dordrecht, 2000.
34. https://youtu.be/2iSE6XlFXhA (accessed February 2007).
35. https://youtu.be/dQhhcgn8YZo (accessed July 2018).
36. https://youtu.be/IGjCOJqINPA (accessed June 2018).
37. L. Feng, S. Li, Y. Li, H. Li, L. Zhang, J. Zhai and D. Zhu, *Adv. Mater.*, 2002, **14**(24), 1857.
38. C. Neinhuis and W. Barthlott, *Ann. Bot.*, 1997, **79**(6), 667.
39. P. Ball, *Nature*, 1999, **400**(6744), 507.
40. J. Zhai, H. J. Li, Y. S. Li, S. H. Li and L. Jiang, *Physics*, 2002, **31**, 483.
41. Y. Liu, J. Tang, R. Wang, H. Lu, L. Li, Y. Kong and J. H. Xin, *J. Mater. Chem.*, 2007, **17**(11), 1071.
42. S. Maharjan, K. S. Liao, A. J. Wang, K. Barton, A. Haldar, N. J. Alley and S. A. Curran, *Mater. Chem. Phys.*, 2020, **239**, 122000.
43. W. Xi, Z. Qiao, C. Zhu, A. Jia and M. Li, *Appl. Surf. Sci.*, 2009, **255**(9), 4836.
44. Q. Pan, H. Jin and H. Wang, *Nanotechnology*, 2007, **18**(35), 355605.
45. R. Blossey, *Nat. Mater.*, 2003, **2**(5), 301.
46. A. Lafuma and D. Quere, *Nat. Mater.*, 2003, **2**(7), 457.

47. L. Feng, S. Li, Y. Li, H. Li, L. Zhang, J. Zhai and D. Zhu, *Adv. Mater.*, 2002, **14**(24), 1857.
48. R. Abu Much, S. Basheer, A. Basheer and M. Hugerat, *J. Chem. Educ.*, 2013, **90**(9), 1207.
49. H. R. Goldsmith, *J. Chem. Educ.*, 2000, **77**, 41.
50. S. Y. Stevens, C. Delgado and J. S. Krajcik, *J. Res. Sci. Teach.*, 2010, **47**(6), 687.
51. A. B. D. Nandiyanto, F. N. Asyahidda, A. A. Danuwijaya, A. G. Abdullah, N. I. A Amelia, M. N. Hudha and M. Aziz, *J. Eng. Sci. Technol.*, 2018, **13**(5), 1352.
52. T. Stuchinskaya, M. Moreno, M. J. Cook, D. R. Edwards and D. A. Russell, *Photochem. Photobiol. Sci.*, 2011, **10**(5), 822.
53. X. Huang, I. H. El-Sayed, W. Qian and M. A. El-Sayed, *J. Am. Chem. Soc.*, 2006, **128**, 2115–2120.
54. Y. Li, H. J. Schluesener and S. Xu, *Gold Bull.*, 2010, **43**(1), 29.
55. S. Ahmed, M. P. N. Bui and A. Abbas, *Biosens. Bioelectron.*, 2016, **77**, 249.
56. B. Helmut and R. M. Richards, *J. Inorg. Chem.*, 2001, **2001**(10), 2455.
57. A. D. McFarland, C. L. Haynes, C. A. Mirkin, R. P. Duyne and H. A. Godwin, *J. Chem. Educ.*, 2004, **81**(4), 544A.
58. K. S. Merza, H. D. Al-Attabi, Z. M. Abbas and H. A. Yusr, *Green Sustainable Chem.*, 2012, **2**(1), 26.
59. J. Popplewell and S. Charles, *New Sci*, 1980, **87**(1220), 332.
60. R. E. Rosensweig, *Sci. Am.*, 1992, **247**(4), 136.
61. A. B. Ellis, M. J. Geselbracht, B. J. Johnson, G. C. Lisensky and W. R. Robinson, *Teaching General Chemistry: A Material Science Companion*, American Chemical Society, Washington, D.C, 1993.
62. J. Dobson, *Drug Dev. Res.*, 2006, **67**(1), 55.
63. S. Ben-David Makhluf, R. Qasem, S. Rubinstein, A. Gedanken and H. Breitbart, *Langmuir*, 2006, **22**(23), 9480.
64. S. B. D. Makhluf, R. Abu-Mukh, S. Rubinstein, H. Breitbart and A. Gedanken, *Small*, 2008, **4**(9), 1453.
65. P. Berger, N. B. Adelman, K. J. Beckman, D. J. Campbell, A. B. Ellis and G. C. Lisensky, *J. Chem. Educ.*, 1999, **76**(7), 943.
66. R. A. Mukh-Qasem and A. Gedanken, *J. Colloid Interface Sci.*, 2005, **284**(2), 489.
67. R. Abu Much and A. Gedanken, *Chem. – Eur. J.*, 2008, **14**(32), 10115.
68. R. Abu Much, A. Meridor and A. Gedanken, *J. Phys. Chem. B*, 2006, **110**, 8194.
69. R. Abu Much and A. Gedanken, *J. Phys. Chem. C*, 2008, **112**, 35.
70. V. Biju, T. Itoh, A. Anas, A. Sujith and M. Ishikawa, *Anal. Bioanal. Chem.*, 2008, **391**(7), 2469.
71. G. O'Neill and T. McMahon, *Emerging Issues in the Practice of University Learning and Teaching*, AISHE, Dublin, 2005, pp. 27–36.

CHAPTER 7

Nanoliposomes as a Model for Teaching Nanochemistry

7.1 Introduction

This chapter presents a teaching module of nanoscience and nanotechnology designed for both secondary school students and undergraduates.

Science teaching curriculums for secondary students around the world include contents on cell structure and its function. Students learn extensively about the cell and its different organelles, including the cell membrane. The main purpose of the proposed teaching module is to optimize the study material so that students learn and understand much more complex organic processes and systems and its impact on daily life. Various studies over the years indicate that many teaching hours with diverse teaching strategies are devoted to the subject of the cell. However, students continue to have difficulty answering questions in biology and they also fail biology tests that deal with these topics.[1]

The module integrates scientific background and laboratory activities which gradually build the connection between the subjects of cell membrane and drug-loaded nanoliposome structures. It consists of two appendices: Appendix 1, consists of scientific background, and Appendix 2 consists of a description all the laboratory activities.

After the application of the presented teaching module, the students learn new concepts of hydrophobic and hydrophilic materials, surfactants, phospholipids, nanoliposome structures and drug delivery.

7.2 Module Structure

The module is built around five meetings (Table 7.1), each meeting lasts between one to four hours. The first meeting focuses on teaching in detail the

Advances in Chemistry Education Series No. 9
Nanochemistry for Chemistry Educators
By Riam Abu Much, Kurt Winkelmann and Muhamad Hugerat

Published by the Royal Society of Chemistry, www.rsc.org

Table 7.1 A description of the proposed teaching module.

Meeting number	Meeting name	Meeting duration (One lesson = 45 minutes. One lecture = 1.5 hours)	Goal	Description
1	Cell membrane and its chemical structure	4 lessons for secondary school students Two lectures for undergraduates	Students will learn and understand the basic scientific concepts: hydrophilicity, hydrophobicity, surfactants, phospholipids, and the structure of cell membrane.	Frontal teaching of the subject of cell membrane and its chemical structure. At this point, the teacher teaches all the relevant scientific background needed (see Appendix 1). The teacher could prepare PowerPoint presentations based on the materials in Appendix 1.
2	Liposome structures: spherical structures of cell membrane	One lesson for secondary school students One lecture for undergraduates	The students learn about chemical structure of liposomes and compare it with the structure of the cell membrane.	Frontal teaching of the subject of liposome structures and its chemical nature. The teaching will be based on the scientific background described in Appendix 1. At the end, the students conclude that liposomes are spherical structures of cell membrane. They could differentiate between the hydrophobic and hydrophilic regions both in cell membrane and liposomes.
3	The amphiphilic structure of phosphatidylcholine	Two lessons for secondary school students One lecture for undergraduates	The students explore the dissolution mode of phosphatidylcholine in water, O/W, and W/O emulsion. In addition, they explore the bilayer structure of liposomes. (Experiments 1–3, Appendix 2).	Laboratory activity (see details in Appendix 2). The students prepare microscale liposomes, O/W, and W/O emulsions, and view them using optical microscope.

Table 7.1 (*Continued*)

Meeting number	Meeting name	Meeting duration (One lesson = 45 minutes. One lecture = 1.5 hours)	Goal	Description
4	Food colors for the identification of hydrophobic and hydrophilic regions	Four lessons for secondary school students	The students detect the hydrophobic and hydrophilic regions of liposomes, O/W, and W/O emulsions, by using two kinds of food colors, hydrophobic and hydrophilic ones. (Experiments 4, Appendix 2).	Laboratory activity.
		Two lectures for undergraduates		The students prepare 6 different solutions (as described in Table 7.2, Appendix 2). The solutions look seemingly the same, but when viewed by an optical microscope the students reveal different structures.
5	Drug-loaded nanoliposomes	Two lessons for secondary school students	The students "mimic" scientists and prepare drug loaded nanoliposomes	Laboratory activity.
		Two lectures for undergraduates		The students prepare microscale liposomes loaded with a colored solution of Optalgin. Then they convert the size of drug-loaded liposomes from microscale to nanoscale and differentiate between them by observing the Tyndall effect.

subject of the cell membrane and its structure, it is considered a basic topic that is taught in secondary schools (usually for Grade Nine or Ten) and in basic scientific courses for undergraduates. It is highly recommended to discuss and highlight basic science concepts, such as hydrophobic and hydrophilic materials, surfactants, and phospholipids. (A detailed scientific background is provided on Appendix 1.) The second meeting offers the possibility to extend the subject of "cell membrane" to the advanced topics of liposome structures, nanoliposomes, and drug delivery. The following last three meetings include three proposed laboratory activities in which the students could prepare different liposome structures and micelles, differentiate between hydrophobic and hydrophilic regions, and finally the students prepare drug-loaded nanoliposome structures.

Appendix 1: Scientific Background

Water is an essential component of life and the most abundant compound on earth. It occupies about 75% of the earth's surface and over 60% of the human body. More than two thousand years ago, Aristotle considered water as one of the main components of all materials.[2]

Chemically, the structure of a water molecule exhibits unique features that make it a very common and effective solvent. Each molecule of water contains two hydrogen atoms bound covalently to one oxygen atom. The covalent bonds between hydrogen and oxygen atoms are formed by sharing a pair of electrons, one electron from the hydrogen atom and the other from the oxygen atom. The electronegativity difference between oxygen and hydrogen is 1.4. This indicates that the covalent bond between them is unequally shared with the electrons bond, oxygen atom has higher tendency to attract electrons than hydrogen atom, this in turn leads to polar covalent bonds between them. Consequently, in a water molecule the hydrogen atoms end up with a partial positive charge and the oxygen atom a partial negative charge.[3]

As two water molecules approach one another, the oxygen atom of one water molecule binds to hydrogen atom of the other water molecule. These bonds are called hydrogen bonds. Generally, hydrogen bonding exists in molecules with a hydrogen atom that is covalently bonded to a very electronegative atom such as oxygen, nitrogen, and fluorine. In comparison to covalent bonds, intermolecular hydrogen bonds are weaker, but they are strong enough to attract water molecules together and thereby confer onto water its unique properties. Figure 7.1 illustrates the bonding phenomenon in water.[3]

Water's molecular polarity and intermolecular hydrogen bonds result in it being known as the "universal solvent". This is due to its ability to dissolve various materials, such as ionic compounds and other polar molecules, by forming electrostatic interactions. Conversely, nonpolar molecules like fats and oils dissolve weakly in water.[2,3]

When comparing interactions of water molecules with other substances and characterizing its chemical structure, substances are categorized into two types – hydrophilic and hydrophobic materials. The linguistic roots of

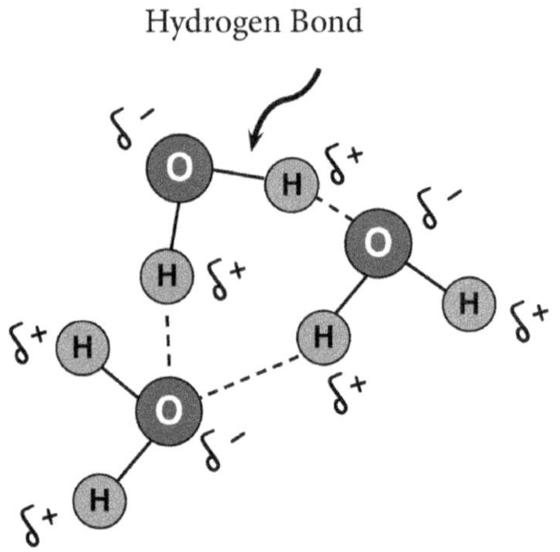

Figure 7.1 Description of bonding phenomenon in water molecule.

hydrophilic translates to "water loving", meaning substances that can interact with water molecules, these substances are polar molecules. Hydrophobic is translated to "water fearing" and these substances do not easily interact with water molecules because of their nonpolar chemical structure. Oils and fats are said to be hydrophobic materials. Another example of hydrophobic materials are candles that are made of paraffin wax, which chemically consists of a mixture of different hydrocarbon chains.[4]

Other types of waxes consist of ester groups formed by binding fatty acid carboxylic groups and hydroxyl groups from long chain alcohols. On the other hand, table sugar, also known as sucrose, is highly soluble in water and it is known to be hydrophilic. The chemical formula of sucrose is $C_{12}H_{22}O_{11}$. Each molecule consists of 12 carbon atoms, 22 hydrogen atoms, and 11 oxygen atoms. Sucrose is a polar molecule that consists of multiple hydroxyl groups and it interacts with water molecules by forming hydrogen bonds.[4]

Why is soap used for cleaning? The answer is related to its chemical structure as soap belongs to a family of materials called surfactants, surface-active agents, or amphiphilic substances. As shown in Figure 7.2, these types of materials have two different sides in their chemical structure. One side is hydrophilic and consists of ionic or polar groups. The other side is hydrophobic and consists of nonpolar groups, usually a hydrocarbon chain. The amphiphilic structure of surfactants like soaps gives them the ability to solubilize both hydrophobic and hydrophilic materials.[5]

When mixed with water, surfactant molecules tend to cluster together in such a way that keeps their hydrophobic sides away from water molecules. At the same time, the hydrophilic side of surfactants are attracted towards the water molecules. Consequently, these dual hydrophobic and hydrophilic interactions of water and surfactant molecules create unique shapes called micelles.[3,5]

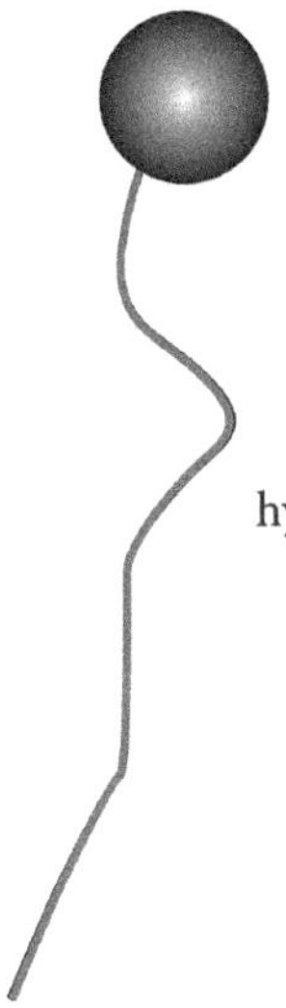

Figure 7.2 Schematic description of surfactant (surface active agent) molecule.

Owing to their chemical structure, water and oil do not mix. Two separate phases are produced when they come in contact with each other. However, scientists succeeded in mixing oil and water by using surfactant molecules. The unique chemical structure of surfactants enables them to disperse oil in water by creating so-called oil-in-water micelles (O/W), where oil molecules are surrounded by the hydrophobic sides of the surfactant molecules. Or, inversely, dispersing water in oil and creating water-in-oil micelles (W/O). These solutions are called emulsions as illustrated in Figure 7.3.[3]

When the size of micelles is at the nanoscale, emulsions are said to be a kind of colloid solution where nanoparticles of materials are dispersed stably in a dispersion medium. In this case, the water phase acts as the dispersion medium. Various studies have demonstrated the great importance of micelles at the nanoscale, especially for medical applications. For example, O/W micelles have been demonstrated to deliver hydrophobic drugs.[6–9]

From the Cell Membrane to Drug Delivery Based on Nanostructures

A biological cell is the smallest, basic unit of life that is responsible for all of life's processes. In principle, each cell contains a fluid called the cytoplasm, which is surrounded by a membrane. In the cytoplasm there also exists many biomolecules such as proteins, nucleic acids, and fats. In addition, cellular structures called cell organelles are also suspended in the cytoplasm.[10]

Cells are made up of multiple cellular organelles that perform specialized functions to carry out life processes. Each organelle has a specific structure. The genetic material of organisms is also present in living cells.

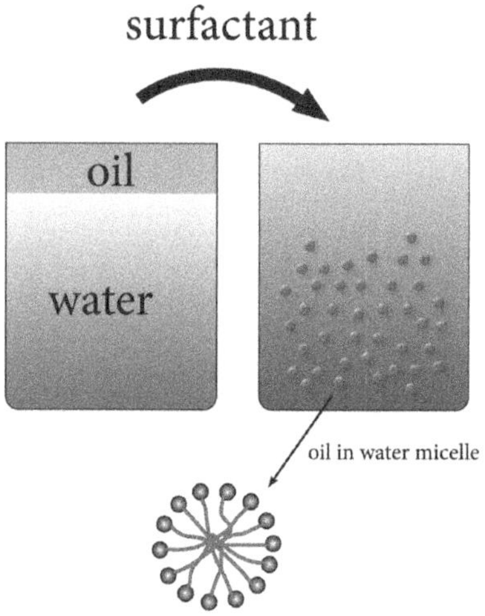

Figure 7.3 Production of emulsions, making oil and water mix.

Each cell that can sustain life executes the following processes:[10]

- Metabolism – this process includes absorption of substances; assembly and decomposition of materials; and excretion of waste.
- Sensitivity and response to stimuli
- Growth and development
- Reproduction and transmission of genetic traits

The basic organelles of biological cells include:

- Cytoplasm – cytoplasm fills the cell volume and contains enzymes that perform metabolic processes. Cytoplasm is where the first stage of cellular respiration takes place. In addition, the cytoplasm carries substances from one place to another inside the cell.
- Cell nucleus – the central place which contains and transfers cell genetic materials.
- Mitochondria – an organelle that releases energy from food to power the cell's biochemical reactions.
- Ribosomes – organelles found in the cytoplasm where proteins are created according to DNA instructions.
- DNA – genetic material located in the cytoplasm in prokaryotic cells (cells without a nucleus) or DNA is surrounded by a cell membrane if located in a eukaryotic cell.

Cell Membrane

All cells are surrounded by a layer called the cell membrane. It consists of a lipid bilayer and various proteins. The cell membrane separates the cytoplasm from the extracellular environment and allows only some substances to pass in and out. In this way, the membrane allows the cell to maintain different conditions inside *versus* outside it and it allows the cell to function as an independent unit. The cell membrane has a liquid character, and it is usually asymmetrical: these properties are crucial for the full functioning of the cell. The cell membrane plays an essential role for cell existence and these roles are described further in the following points:

- It creates a cell boundary from the external environment.
- It regulates the movement of materials between the cell and its surrounding environment.
- It communicates between the cell and other cells.[10–12]

Chemical Structure of Cell Membrane. The cell membrane is composed mainly of lipids into which proteins are incorporated. Most cell membrane lipids are phospholipids. Phospholipids are a type of surfactant that have a hydrophilic head and two hydrophobic fatty acid tails. An illustration of the cell membrane is shown in Figure 7.4. The phospholipids in a cell membrane are arranged in a double layer "sandwich model". One of the sandwich phospholipid heads faces inwards, towards the inner cell cytoplasm and the other phospholipid heads face outwards, towards the outer extracellular fluid environment. Both phospholipid tails face inwards towards the inner part of the membrane. This structure of the cell membrane forms a hydrophobic region between the two aqueous solutions of the cell interior and the cell exterior environment. This allows for selective passage of substances through the membrane as well as maintaining the cell shape despite its fluidity. The two-layer structure minimizes the exposure of the hydrophobic groups to water, so the membrane structure is energetically desirable.[13]

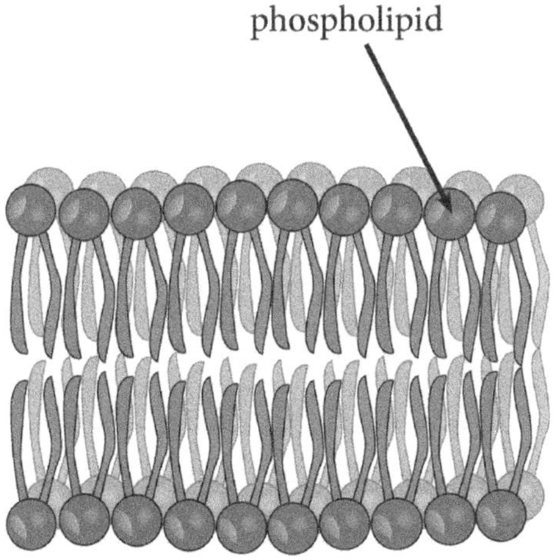

Figure 7.4 Schematic illustration of cell membrane structure.

Chemical Structure and Biological Function of Lipids and Phos-pholipids[14]. Lipids are organic compounds. Each biological molecule that is soluble in oil is a lipid. Oils, fats, vitamins, and cholesterol are all lipids. In living organisms, lipids are one of the basic cell membrane components. The chemical structure of lipids is categorized into several groups. The main categories of them are described below.

- Fatty acids – a type of carboxylic acid than consists of up to 36 carbon atoms. They contain carboxylic groups (–COOH) and long hydrocarbon chains. Fatty acids are divided into two main groups, saturated fatty acids and unsaturated fatty acids. Saturated fatty acids have singular covalent bonds between carbon atoms, and they are saturated with hydrogen atoms. Unsaturated fatty acids have at least one double bond between the carbon atoms.
- Glycerides – glycerides are chemical structures that are composed of glycerol (an alcohol) bonded to fatty acids. Glycerol is a viscous, colourless, and water-soluble liquid. When glycerol is bonded to one fatty acid it is considered a monoglyceride. When it is bonded to two fatty acids it is called a diglyceride. A triglyceride is formed when glycerol is bonded to three fatty acid molecules. The glycerol molecule is composed of a chain of three carbon atoms. Each carbon atom is bonded to a hydrogen atom (one or two) and a hydroxyl group (–OH). Because of the hydroxyl group glycerol belongs to the family of alcohols. Figure 7.5 illustrates the chemical structure of glycerol.
- Phospholipids – these are diglycerides with a hydrophilic group consisting of phosphate and a polar group. The phosphate group is a derivative of phosphoric acid and this group is negatively charged with a physiological pH of 7.

A phospholipid molecule is formed when two hydroxyl groups in glycerol bond to the carboxylic group of fatty acid. The third hydroxyl group of glycerol bonds to a phosphorus group. The resulting molecule is called phosphatide. The phosphorus group in phosphatide is bonded with the hydroxyl group of one of the following molecules: serine, ethanolamine, choline, inositol or glycerol (Figure 7.6).

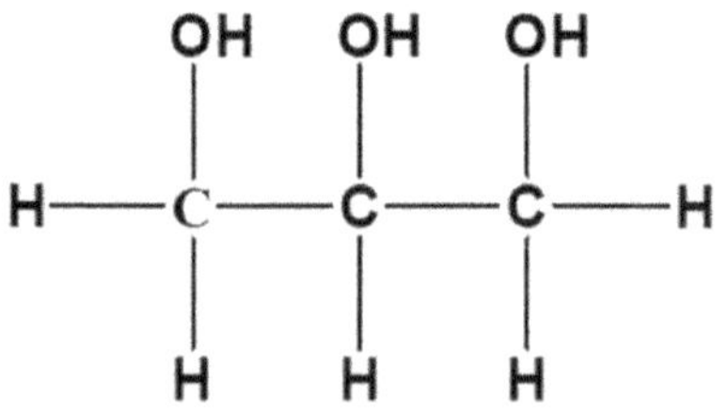

Figure 7.5 Chemical structure of glycerol.

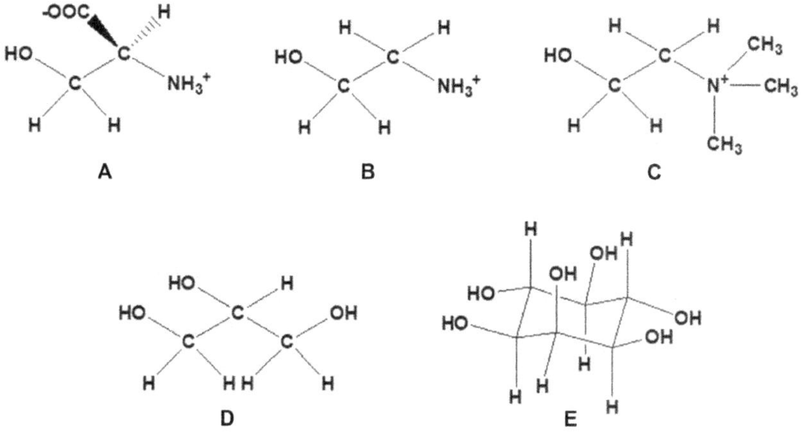

Figure 7.6 Chemical structure of the amino acids, serine, ethanolamine, choline, inositol, and glycerol.

Accordingly, phospholipids contain four basic components: fatty acids, glycerol, a phosphorus group and another chemical group. In nature, the fatty acid hydrocarbon chains in phosphoglycerates usually contain an even number of carbon atoms – usually between 14 to 24.

The most well-known type of phospholipid is phosphatidylcholine. It is a type of phospholipid composed of choline in the head group and two fatty acyl groups as the tail groups. It is the main component of biological membranes, and using a hexane extraction method it can easily be obtained from a variety of sources, such as egg yolk, mustard, sunflower, and soybeans. Figure 7.7 illustrates the chemical structure of phosphatidylcholine.[3]

Phosphatidylcholine is a precursor molecule in producing acetylcholine in the body. It is also considered a brain activity stimulator. Medical studies have shown that eating food supplements containing phosphatidylcholine significantly improves memory, concentration, and learning. In addition, it has been found that it improves disorders related to depression and stress. It was also found that the phosphatidylcholine molecule could significantly improve the quality of life of patients with senility and Alzheimer's Disease. Important advantages of using phosphatidylcholine are that it is considered safe and it does not have reported side effects.[15]

Phospholipid molecules are of paramount importance in nature as they are a major component in all living organism cell membranes. In the human body they constitute the membrane that surrounds the cells and the membrane that surrounds the various organelles in the cell. The unique properties of the biological membrane are due to the phospholipid chemical composition. For example, membrane fluidity depends on the length and degree of fatty acid saturation in the phospholipid molecules. In fact, the phosphatidylcholine molecule is the most common phospholipid molecule in the membrane of our body cells.

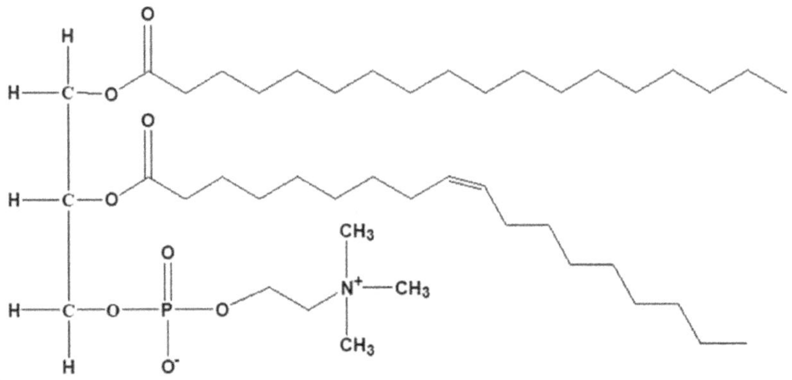

Figure 7.7 Chemical structure of phosphatidylcholine.

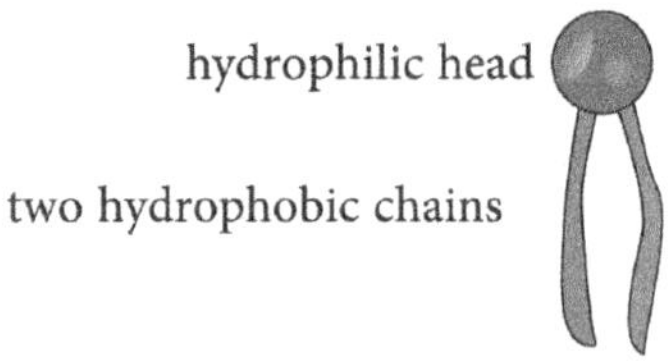

Figure 7.8 Schematic illustration of phospholipid structure of polar head group and two hydrophobic tails.

All types of phospholipids that make up the cell membrane have a common basic structure that categorize them as amphipathic molecules (surfactants). As shown in Figure 7.8, each molecule of phospholipid consists of hydrophilic head that composed mainly of negatively charged phosphate group, and two hydrophobic tails composed of fatty acids hydrocarbon chains.

The bilayer structure of a cell membrane is based on the chemical nature of the different components in the human body. Blood is an aqueous solution of different ionic and polar molecules. So, the phospholipids that make up the cell membrane are arranged in such a way that their hydrophilic part is oriented towards the exterior part of the cell, towards the blood system, and its hydrophobic tails are oriented towards the interior part of the cell membrane.

Phospholipids can be obtained from natural components. For example, phosphatidylcholine can be easily extracted from egg yolks or soybean.

Liposome Structures

The chemical structure of a phospholipid molecule consists of two different parts: the hydrophilic head group and two hydrophobic tails of long hydrocarbon chains. The dissolution mode of phospholipids in water mimics the structure of a cell membrane. It is organized in such a way that phospholipid hydrophobic tails are oriented far from water molecules, whereas the hydrophilic head is oriented towards the water molecule. This results in the formation of bi-layer structures called liposomes. Liposomes can be defined as microscopic vesicles wherein their aqueous volume is surrounded by phospholipid molecules. The name liposome is derived from two Greek words, "lipos" meaning fat, and "soma" meaning body.[16]

Liposomes Mimic Cell Membrane Structure. Comparing between the cell membrane structure and the structure of liposomes reveals great similarities. Liposomes consist of an aqueous core surrounded by a lipid bilayer. Similarly, cell membranes separate a cell's inner aqueous core from the bulk outside. In other words, liposomes are circular structures of the cell membrane. Figure 7.9 illustrates schematically the structure of liposome and its similarity to the structure of cell membrane.

The unique resemblance between the natural structure of cell membranes and liposomes created a breakthrough for medical uses of liposomes such as their use as drug carrier candidates.[17–19] Drug delivery is a method or technological process where drug compounds are intended to act upon a target site under defined conditions. This is in order to achieve an effective therapeutic effect in humans or animals while reducing side effects. The main goal of a drug delivery system is to direct the drug to the appropriate target and thus optimize the drug's interaction with the diseased tissue.

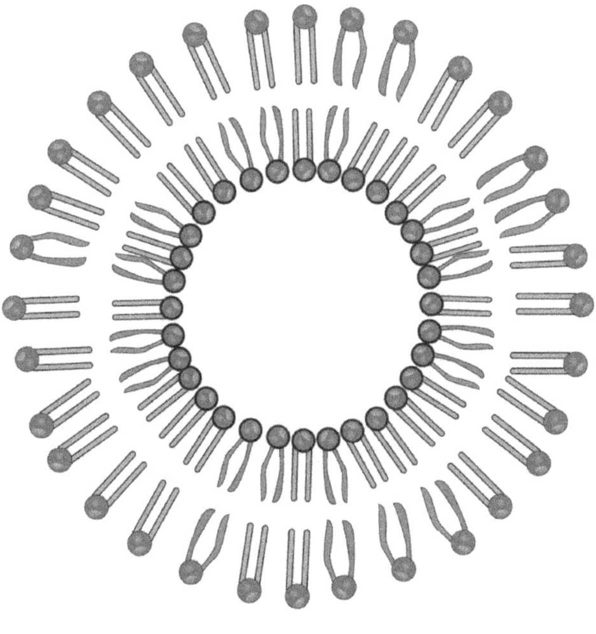

Figure 7.9 Schematic illustration of liposome structures.

Traditional drug delivery, such as oral drug administration or intravenous injection, is based on the drug's absorption into all body tissues so that only a small portion of the drug reaches the target organ. The drug delivery method is based on targeted and timed drug release at the target site. Liposomes are the most frequently used carrier to dispense drugs. Liposomes can be synthetically produced and consist of a double layer of phospholipids that enclose into a circle and form a spherical shape. The liposome can store the drug in the hydrophobic envelope or inside the hydrophilic inner region depending on the properties of the drug being delivered. When the liposome reaches the target site it will join onto the cell membrane and transfer its contents specifically into the cytoplasm of the target cell. In addition to their use as drug carriers, they have multiple uses in cosmetics and food manufacturers for encapsulating unstable compounds such as antimicrobials, antioxidants, and flavorings. Liposomes can carry both hydrophilic and hydrophobic compounds. They are characterized by their biocompatibility, biodegradability, low toxicity, and the ability to carry hydrophobic and hydrophilic drugs. Due to their versatile properties, there has been enhanced interest and research into liposomal structures and their use as drug carriers.[20–22]

Nanoliposomes as a Drug Carrier. The combination of nanotechnology and medicine paved the way for the beginning of a new field called nanomedicine. Nanomedicine includes a wide range of nanotechnological applications, such as drug delivery systems, sensors, fluorescent markers for imaging purposes, and more.[23,24] The nano-revolution in medicine has advanced the development of nano-systems that can carry drug molecules in our body. Due to their tiny size, nano-carrying systems can locate and carry a drug to a specific area in the body. Nano-systems for carrying drugs consist mainly of a carrier material and an active substance. The carrier material serves as the body of the nanoparticle and the active substance targets molecules that allow the nanoparticle to find certain cells in the body. There are now a wide variety of nanoscale carrying systems that differ from each other in various respects. They differ in the materials of which they are composed, in their size, and by the amount of material they can carry.

Nanoliposomes are very attractive as drug delivery vehicles. Nanoliposomes are nanoscale-sized, spherical structures that consist of a circular membrane of phospholipids and are filled with fluid. The nanoliposome acts as a form of nanoscale container that can carry the drug inside it and bring it to its destination. A great example of the use of nanoliposomes as a carrier system is the drug Doxil, which uses a nanoliposome carrier system to treat breast cancer. The active substance in Doxil is doxorubicin. Doxorubicin is located inside the nanoliposome. It is carried towards the tumor in a targeted fashion so that its distribution into the rest of the body and its side effects are significantly reduced. Figure 7.10 illustrates a drug loaded nanoliposome when it penetrates the cell membrane.[25–27]

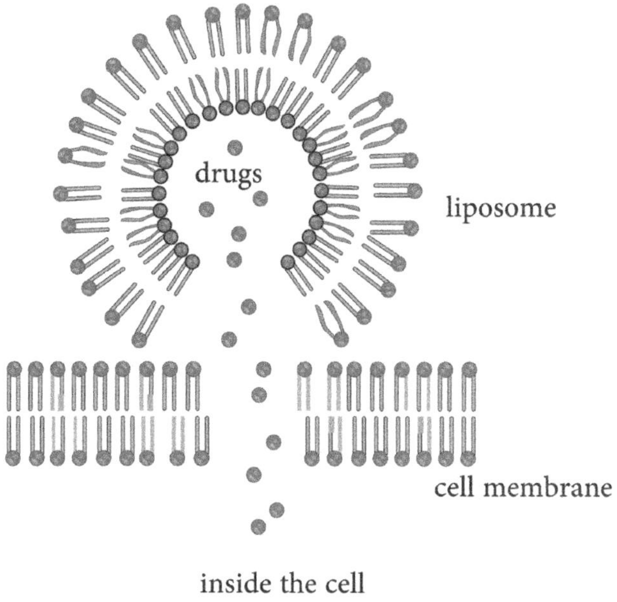

Figure 7.10 Schematic description of drug loaded liposome and its penetration into the cell membrane.

Appendix 2: Detailed Description of Laboratory Activities

The presented teaching module (Table 7.1) consists of laboratory activities that aim to expose students to one of the main medical applications of nanostructured materials. It is intended to be taught to secondary school students and undergraduates. The activities consist of presenting several consecutive steps, starting from introducing the cell membrane chemical structure through to using nanoliposomes as a drug vehicle.[3,28]

The accepted model for the structure of cell membrane is the bi-layer model of lipids and proteins.[29] According to this model, the membrane separates the extracellular environment from the intracellular one. The aim of the presented lab activities is to have students fabricate artificial cell membranes (nanoliposomes) that can mimic the structure of cellular membranes using natural phosphatidylcholine. During the lab activities the students complete the following tasks.

- They produce liposomal structures using the surfactant phosphatidylcholine. Phosphatidylcholine is a natural type of phospholipid which can be extracted from egg yolks or soybean oil.
- They will investigate the ideas of hydrophilicity and hydrophobicity by exploring the dissolution mode of hydrophilic and hydrophobic food colors in liposomal structures.
- They will form a demonstrative model of drug loaded liposomal structures.
- They will reduce the liposome size from the microscale to nanoscale.

Experiment One: The Amphiphilic Structure of Phosphatidylcholine

Phosphatidylcholine is a surfactant which has both hydrophilic and hydrophobic regions in its molecular structure. It dissolves and forms liposomal structures in water. It achieves this by arranging its polar regions towards the water molecules and its non-polar tails away from the aqueous phase.[3]

In this experiment, students explore the dissolution mode of phosphatidylcholine in three environments: pure water, oil in water emulsion, and water in oil emulsion.[3] This experiment could be suitable for secondary school students and undergraduates and, accordingly, two experimental descriptions for each age group are described below.

Materials needed:

- Phosphatidylcholine
- Double distilled water
- Vegetable oil
- iso-Propanol

Experimental Instructions for Undergraduate Students[3].

1. In a 50 mL round bottom flask and while continually stirring, add 0.5 g of phosphatidylcholine into 25 mL of iso-propanol.
2. Stir for about two minutes in order to dissolve the phosphatidylcholine.
3. Evaporate the solution under vacuum at 50 °C in order to remove the solvent and to allow the lipids to form a film onto the surface of the flask.
4. Finally, dissolve the lipid film into 100 mL of double-distilled water *via* stirring with a magnetic stir bar for one hour.

Experimental Instructions for Secondary School Students.

1. Simply insert a small amount of phosphatidylcholine into a plastic tube.
2. Add double-distilled water into the tube.
3. Shake the tube vigorously using a test tube vortex shaker.

In these experimental set-ups, students can witness the organizational mode of phospholipid molecules in water, and they can view the liposome structures under an optical microscope. Figure 7.11 shows how students will be able to differentiate between the inner hydrophilic region of liposomes and the hydrophobic ring that separates the interior and exterior hydrophilic regions. When dissolving phosphatidylcholine in water, it organizes itself in such a way as to keep its hydrophobic chains away from the water phase. According to Riam *et al.* microscale liposome structures are formed.[3,28] In addition, students can compare between a cell membrane and liposomal

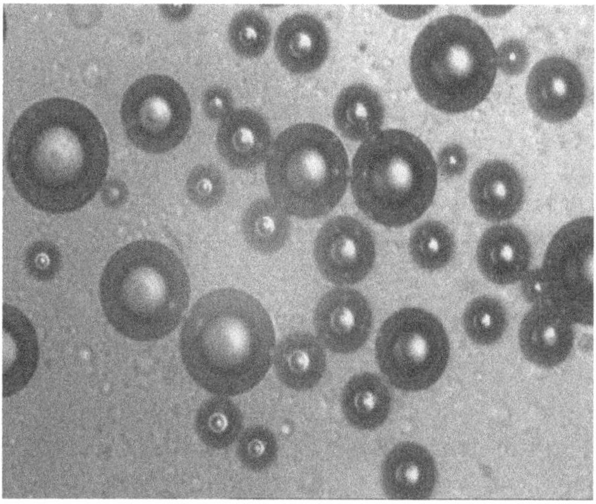

Figure 7.11 An image of microscale liposome structures obtained by optical microscope view (×1000).

structures and discover that liposomes are considered a circular structure of a cell membrane.

Water and oil do not dissolve into each other because of their different chemical structures. Van der Waals interactions are a type of intermolecular interactions between oil molecules while water molecules interact with each other *via* hydrogen bonding. Therefore, oil and water are insoluble within each other and are called immiscible liquids. Consequently, when trying to mix them together, two separate phases are formed. The addition of phosphatidylcholine causes the two phases of oil and water to be mixed into one phase, called an emulsion. When using an optical microscope, the students can view the resulted structures called micelles. They can further explore the role of phosphatidylcholine upon mixing two immiscible liquids.

When attempting to dissolve oil into water, the emulsion produced is called an O/W emulsion. Upon adding phosphatidylcholine, it self-organizes in such a way that the phosphatidylcholine hydrophobic chains become tightly packed to avoid the water molecules. Oil molecules can then be entrapped inside the hydrophobic region of the micelles. In contrast, when trying to dissolve water into oil, the phosphatidylcholine organizes in a reverse fashion. Its hydrophobic chains orient towards the oil molecules and its hydrophilic heads trap the water molecules to keep them away from the oil molecules.

Experiment Two: Preparation of an O/W Emulsion using Phosphatidylcholine

1. In a chemical beaker add 1.5 g vegetable oil, 0.3 g phosphatidylcholine, and 30 mL double-distilled water.
2. Mix the solution using a magnetic stirrer for one hour.

Experiment Three: Preparation of a W/O Emulsion using Phosphatidylcholine

1. In a chemical beaker, add 1.5 mL double-distilled water, 0.3 g phosphatidylcholine, and 30 mL vegetable oil.
2. Mix the solution using magnetic stirrer for one hour.

Experiment Four: Identification of Hydrophilic and Hydrophobic Regions

According to a study by Riam *et al.*, the authors propose a lab activity which enables students to differentiate between the hydrophobic and hydrophilic regions of micro-scale micelles and liposomal structures.[3] The lab experiment is based on simply using hydrophilic and hydrophobic food dyes during the preparation of both micelles and liposomes. For this purpose, it is recommended that students be asked to prepare six solutions with the following compositions as displayed in Table 7.2.

In solution number one, the students explore the dissolution mode of phosphatidylcholine in an aqueous solution of hydrophilic food dye. As described in the previous section, liposomal structures are produced in this case. In solution number two, students produce liposomal structures in water with the addition of hydrophobic food dye. In the rest of the solutions (solutions number three through six), the students prepare emulsions with different compositions. In the third and fourth solutions they prepare oil-in-water liposomes with hydrophilic and hydrophobic food dyes, respectively. In the fifth and sixth solutions, water-in-oil emulsions are created with hydrophilic and hydrophobic food dyes, respectively.

The following materials are needed for the proposed lab experiments described below.

- Phosphatidylcholine
- Double distilled water
- Vegetable oil
- iso-Propanol
- Hydrophilic red food dye
- Hydrophobic red food dye

Dissolving Phosphatidylcholine in an Aqueous Solution of Hydrophilic Food Dye (Solution 1 in Table 7.2)

Experimental Instructions for Undergraduate Students.

1. Pour 25 mL of iso-propanol into a 50 mL round bottom flask.
2. Weigh 0.3 g of phosphatidylcholine.
3. Add phosphatidylcholine into the 25 mL of iso-propanol.
4. Stir for about two minutes in order to dissolve the phosphatidylcholine.

Table 7.2 Components of six solutions prepared.

Number of solution	Quantity of hydrophobic red dye/mg	Quantity of hydrophilic red dye/mg	Quantity of phosphatidylcholine/g	Quantity of vegetable oil/mL	Quantity of double distilled water/mL
1	0	10	0.3	0	100
2	10	0	0.3	0	100
3	0	10	0.3	5	95
4	10	0	0.3	5	95
5	0	10	0.3	95	5
6	10	0	0.3	95	5

5. Evaporate the solution under vacuum at 50 °C in order to remove the solvent and to allow the lipids to form a film onto the surface of the flask.
6. In a different flask, prepare an aqueous solution by dissolving 10 mg of hydrophilic food dye into 100 mL double-distilled water.
7. Dissolve the lipid film into the aqueous solution of hydrophilic food dye *via* stirring with a magnetic stir bar for one hour.

Experimental Instructions for Secondary School Students.

1. Insert 0.3 g of phosphatidylcholine into a plastic tube.
2. Add 10 mg of hydrophilic food dye.
3. Add 100 mL double-distilled water into the tube.
4. Shake the tube vigorously using a test tube vortex shaker.

Dissolving Phosphatidylcholine in Double-Distilled Water with Hydrophobic Food Dye (Solution 2 in Table 7.2)

Experimental Instructions for Undergraduate Students.

1. Pour 25 mL of iso-propanol into a 50 mL round bottom flask.
2. Weigh 0.3 g of phosphatidylcholine.
3. Add phosphatidylcholine into the 25 mL of iso-propanol.
4. Stir for about two minutes in order to dissolve the phosphatidylcholine.
5. Evaporate the solution under vacuum at 50 °C in order to remove the solvent and to allow the lipids to form a film onto the surface of the flask.
6. In a different flask, prepare an aqueous solution by dissolving 10 mg of hydrophobic food dye into 100 mL double-distilled water.
7. Dissolve the lipid film into the aqueous solution of hydrophobic food dye *via* stirring with a magnetic stir bar for one hour.

Experimental Instructions for Secondary School Students.

1. Insert 0.3 g of phosphatidylcholine into a plastic tube.
2. Add 10 mg of hydrophobic food dye.
3. Add 100 mL double-distilled water into the tube.
4. Shake the tube vigorously using a test tube vortex shaker.

Solution 3: O/W Emulsion in the Presence of Hydrophilic Food Dye

1. In a chemical beaker add 5 mL vegetable oil, 95 mL double-distilled water, 0.3 g phosphatidylcholine, and 10 mg hydrophilic food dye.
2. Mix the solution using a magnetic stirrer for one hour.

Solution 4: O/W Emulsion in the Presence of Hydrophobic Food Dye

1. In a chemical beaker add 5 mL vegetable oil, 95 mL double-distilled water, 0.3 g phosphatidylcholine, and 10 mg hydrophobic food dye.
2. Mix the solution using a magnetic stirrer for one hour.

Solution 5: W/O Emulsion in the Presence of Hydrophilic Food Dye

1. In a chemical beaker add 95 mL vegetable oil, 5 mL double-distilled water, 0.3 g phosphatidylcholine, and 10 mg hydrophilic food dye.
2. Mix the solution using a magnetic stirrer for one hour.

Solution 6: W/O Emulsion in the Presence of Hydrophobic Food Dye

1. In a chemical beaker add 95 mL vegetable oil, 5 mL double-distilled water, 0.3 g phosphatidylcholine, and 10 mg hydrophobic food dye.
2. Mix the solution using a magnetic stirrer for one hour.

According to Riam *et al.*, despite their different compositions, the six solutions described above look like the same red-colored solutions.[3] The students discover the differences in their composition by viewing them under an optical microscope. Figure 7.12 presents example images which can be obtained from these solutions. Figure 7.12a presents liposomal

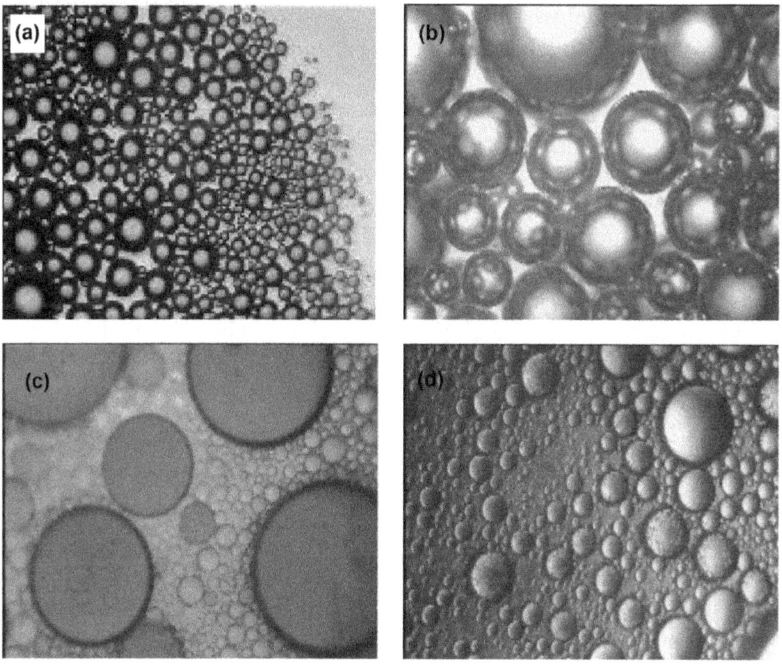

Figure 7.12 Images obtained by optical microscope (×1000) of: a, liposomes prepared with hydrophilic red dye, b, liposomes prepared with hydrophobic red dye, c, oil-in-water micelles prepared with hydrophobic dye, d, oil-in-water micelles prepared with hydrophilic dye.

structures obtained from solution number 1. In this image the students can visualize that the interior hydrophilic region of the liposomes is red-colored which is the characteristic color of the applied, red hydrophilic food dye. In addition, the liposome structures are dispersed in the red-stained aqueous solution.

When liposomes are prepared with the addition of a hydrophobic food color, no dye is detected either in the interior region of the liposomes nor in the aqueous solution. The food color is trapped in the hydrophobic ring of the liposomes, as shown in Figure 7.12b. Different micelles are formed in this situation, and the colored, spherical micelles that are dispersed in the non-colored solution can be detected. They can be detected with oil-in-water micelles prepared with hydrophobic dye (Figure 7.12c) and oil-in-water micelles prepared with hydrophilic dye. In contrast, non-colored spherical micelles dispersed in colored solutions can be detected when preparing an oil-in-water solution with a hydrophilic food color and water-in-oil solution with a hydrophobic food color (Figure 7.12d).

Experiment 5: Preparation of Microscale Drug Loaded Liposome Structures

Based on the results of experiment number 4, students can potentially conclude that any material could penetrate different regions inside the liposomal structures depending on its chemical structure and whether it is hydrophobic or hydrophilic. These features of liposomes make them useful candidates for delivering different kinds of drugs. In addition, liposomal structural similarities to the chemical structure of cell membranes enables them to easily penetrate the cells in order to inject a drug.

During this experiment, students can choose tablets from drug types such as Optalgin, Augmentin, or vitamin C. They must then grind them and dissolve them into double-distilled water. In order to identify the drug inside the liposomal structures, the aqueous solution is dyed with hydrophilic food dye. Liposomes are then prepared by dissolving phosphatidylcholine into the dyed, aqueous drug solution. Exact steps of this experiment are described below.

Experimental Instructions for Undergraduate Students

1. Grind one Optalgin tablet using a griding mill.
2. Prepare a saturated solution by dissolving the ground Optalgin into 10 mL double-distilled water.
3. Add 10 mg red hydrophilic dye to the saturated solution in order to color the solution.
4. Weigh 0.3 g of phosphatidylcholine.
5. Add phosphatidylcholine into the 25 mL of iso-propanol.
6. Stir for about two minutes in order to dissolve the phosphatidylcholine.

7. Evaporate the solution under vacuum at 50 °C in order to remove the solvent and to allow the lipids to form a film onto the surface of the flask.
8. In a different flask, prepare an aqueous solution by the dyed Optalgin solution into 90 mL double-distilled water.
9. Dissolve the lipid film into the aqueous solution of Optalgin solution prepared in step 8.

Experimental Instructions for Secondary School Students

1. Grind one Optalgin tablet using a griding mill.
2. Prepare a saturated solution by dissolving the ground Optalgin into 10 mL double-distilled water.
3. Add 10 mg red hydrophilic dye to the saturated solution in order to color the solution.
4. Inside a plastic tube, mix 0.3 g of phosphatidylcholine into 90 mL of double-distilled water.
5. Add the Optalgin solution into the plastic tube.
6. Shake vigorously using a test tube vortex shaker.

In the experiments above, by using an optical microscope, students can visually explore the position of the drug inside the liposomes. During the preparation of liposomes inside an aqueous drug solution, some of the dyed drug penetrates the inner hydrophilic region of the liposomal structures. The rest of the drug remains dispersed in the aqueous solution medium.

Drug Loaded Liposome Structures: From Microscale to Nanoscale

The final activity of the suggested lab experiments is designed for students to convert drug loaded liposomes from microscale to nanoscale by using an Avanti® Mini-Extruder apparatus. The Avanti® Mini-Extruder is a simple device that reduces the size of liposomal structures to nanoscale dimensions in a simple and effective way.[3] The activity is based on passing the liposomal solution through a nanoscale sized, porous membrane made of polycarbonates. There are different types of membranes which differ in the size of their pores. Some range from 50 nm to 1000 nm in pore size. Figure 7.13 presents a depiction of the Avanti® Mini-Extruder.

How to Compare Between Microscale and Nanoscale Sized Liposomes

The Tyndall effect is considered an easy, simple, and visual method for the detection of nanoscale structures in solution. The Tyndall effect is also known as "Tyndal scattering". It is a phenomenon of light scattering by particles in a colloidal solution or suspension. Typically, this effect is used to

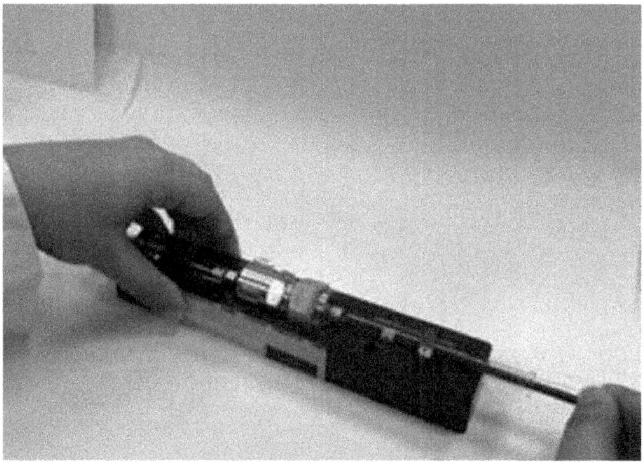

Figure 7.13 An image of Avanti Mini-Extruder apparatus.

differentiate between a true solution, colloidal solution (a solution of nanoscale particles), and a suspension (a solution of microscale particles).

Accordingly, there are two main effects that may occur when shining a beam of light through a solution. The first effect shows the beam passing through the solution in straight lines, without scattering. The second option displays how the light scatters randomly into different directions due to its collision with solute particles. Both types of effects depend on the size of the solute particles. In a true solution, the solute particle sizes are generally bellow 1 nm. In this case the light passes through the solution without any interference because of the small size of solute particles and therefore no scattering occurs. However, when passing a beam of light through microscale- or nanoscale-sized particles, the light collides with the solute particles and scattering occurs. Visually, a beam of light is detected through the solution.

References

1. T. Elangovan and Z. Ismail, *Education*, 2013, **3**(4), 231.
2. S. S. Zumdhal and A. A. Zumdhal, *Chemistry*, Brooks Cole, Belmont, USA, 2014.
3. R. Abu Much, S. Basheer, A. Basheer and M. Hugerat, *J. Chem. Educ.*, 2013, **90**(9), 1207.
4. F. Talal, A. Mohammed, O. Ibrahim and A. D. Mohammed, *Sci. Rep.*, 2017, 7(1), 7531.
5. P. Juan and M. Martin, *J. Chem. Educ.*, 2011, **88**(5), 629.
6. M. J. Rosen, *Surfactants and Interfacial Phenomena*, John Wiley & Sons, New York, 1989.
7. S. Mall, G. Buckton and D. A. Rawlins, *J. Pharm. Sci.*, 1996, **85**(1), 75.

8. J. Casado, C. Izquierdo, S. Fuentes and M. L. Moya, *J. Chem. Educ.*, 1994, **71**, 446.
9. E. M. Ebeid, *J. Chem. Educ.*, 1985, **62**(2), 165.
10. C. A. Kerfeld, M. R. Sawaya, S. Tanaka, C. V. Nguyen, M. Phillips, M. Beeby and T. O. Yeates, *Science*, 2005, **309**(5736), 936.
11. M. Edidin, *Nat. Rev. Mol.*, 2003, **4**(5), 414.
12. H. T. McMahon and J. L. Gallop, *Nature*, 2005, **438**(7068), 590.
13. M. S. Bretscher, *Sci. Am.*, 1985, **253**(4), 100.
14. C. K. Mathwes, K. E. VanHolde and K. G. Ahern, *Biochemistry*, Pearson, United Kingdom, 1999.
15. F. Chen, Q. Zhao, X. Cai, L. Lv, W. Lin, X. Yu and X. G. Wang, *Can. J. Microbiol.*, 2009, **55**(11), 1328.
16. R. Yusaf, R. Nawaz, S. Hayat, A. Khursheed, N. Zafar, A. Ahmad and I. Majeed, *Indo Am. J. Pharm*, 2014, **4**(8), 3559.
17. C. Wang, J. Piao, Y. Li, X. Tian, Y. Dong and D. Liu, *Angew. Chem. Int.*, 2020, **59**(35), 15176.
18. R. Kamble, V. B. Pokharkar, S. Badde and A. Kumar, *Int. J. Pharm. Pharm. Sci.*, 2010, **2**(4), 87.
19. S. Shailesh, S. Neelam, K. Sandeep and G. D. Gupta, *J. Pharm. Res*, 2009, **2**(7), 1163.
20. N. Bertrand and J. C. Leroux, *J. Control. Release*, 2012, **161**(2), 152.
21. M. R. West, T. W. Hanks and R. T. Watson, *J. Chem. Educ.*, 2009, **86**(3), 373.
22. S. S. Patel, *Pharmainfo. Net*, 2006, **4**(5), 1.
23. A. Moshed, A. Mohammad, M. K. Islam Sarker and M. Khaleque, *Am. J. Biomed. Sci.*, 2017, **9**(1), 1.
24. R. Misra, S. Acharya and S. K. Sahoo, *Drug Discov.*, 2010, **15**(19–20), 842.
25. L. Silverman and Y. Barenholz, *Nanomedicine*, 2015, **11**(7), 1841.
26. S. Hallaj-Nezhadi and M. Hassan, *Drug Deliv.*, 2015, **22**(5), 581.
27. Z. Ali Mohammadi, S. Foad Aghamiri, A. Zarrabi and M. Reza Talaie, *Curr. Drug Deliv.*, 2016, **13**(7), 1065.
28. R. Abu-Much, N. Najami and S. Hugerat, *Int. J. Chem. Sci.*, 2017, **15**(4), 209.
29. S. K. Muhammad, S. D. Noura and D. W. John, *Int. J. Mol. Sci.*, 2013, **14**, 21561.

CHAPTER 8

Social and Ethical Issues of Nanotechnology

8.1 Introduction

Societal issues and ethics may not be an obvious topic for a chemistry classroom. However, all practicing scientists consider the ethical implications of their work. Professional science and engineering societies have codes of ethics. When communicating with the public, chemists and educators have a duty to clearly and accurately express scientific concepts and findings. They must perform their work in a way that ensures the safety of co-workers, customers, students, and the public. Protecting the public includes protecting the environment and mitigating the damage caused by chemical products and waste. Researchers developing pharmaceuticals, personal care products, and similar materials follow strict protocols when testing new chemicals to ensure that they are safe for their intended use. The testing process itself, whether it involves animals or humans, follows rigorous guidelines. Like other social science researchers, educators studying ways to improve student learning ensure that their new pedagogies or curriculum materials at a minimum do not harm their students. Some of these ethical guidelines are enforced by law and professional certification. Others, such as Green Chemistry (Section 5.2.4.2) are aspirational but not required.

Scientists now have a better understanding of the impact of the chemical industry on society. They are more aware of how the benefits and risks are unevenly distributed among different groups within society. For instance, traditionally marginalized peoples often lack the political power to influence policy decisions and therefore they tend to suffer more negative consequences but reap fewer benefits. This is also true regarding the quality of science education that students can access. New knowledge leads chemists and

Advances in Chemistry Education Series No. 9
Nanochemistry for Chemistry Educators
By Riam Abu Much, Kurt Winkelmann and Muhamad Hugerat

Published by the Royal Society of Chemistry, www.rsc.org

educators to reconsider how they define expectations for ethical behaviour beyond traditional standards of environmental protection, worker safety, and honesty.

It is possible to teach chemistry in a way that it seems unrelated to a student's everyday life. To avoid this, teachers connect chemical concepts to current events, everyday activities, and consumer products that spark students' curiosity about the chemical nature of the world around them. Many students are interested in rectifying social injustices and improving their communities. How societies address unequal distribution of benefits and risks from scientific endeavours is one of many important ethical considerations that students should consider. In doing so, they will realize how important ethics is in chemistry. These justifications for teaching ethics as part of a student's education in chemistry apply to nanotechnology.

Teaching ethics is part of students' overall education as they become well-rounded, socially conscious citizens. In particular, scientific ethics is increasingly relevant because new technologies are more frequently the source of potential problems or a solution to problems. Technical knowledge can contribute to the decision-making process but it alone cannot lead to the best answer. Students should start thinking about the type of world that they wish to live in and how nanotechnology can contribute to its development.

Nanotechnology is a relatively new field so there is a greater need for educating the public, including middle and high school students. Knowledge about nanotechnology is growing among adults in many countries. It has been observed generally that increased awareness correlates with a more favourable opinion of nanotechnology.[1]

Resources are available for teachers wishing to include an ethics activity in their classes, although fewer than are found for laboratory activities. This reflects the limited extent that ethics appears in any science curriculum. Many of the activities referenced here are available from government agencies such as the National Nanotechnology Infrastructure Network (NNIN)[2] in the United States and the Nano for Youth (NANOYOU) website[3] from the European Union. Their availability is due to a concerted effort by policy makers to educate the general population about nanotechnology. Unlike past instances in which the public participated in discussions only at the end of the technology development process, nanotechnology policy makers and scientists have engaged with the public since the initial stages of government funding of nanotechnology research. Previous chapters in this book explain that while nanotechnology builds upon traditional fields of science and engineering, nanotechnology has many unique qualities. These raise concerns, not only due to the unusual chemical and physical properties of nanomaterials but also due to the power of nanotechnology. Future developments in medicine, electronics, military applications, energy, and consumer goods based on nanotechnology can radically change how individuals relate to each other, their government, and the world around them.

8.2 Government Support and Public Engagement

Nanotechnology research and development (R&D) in the United States is supported through the National Nanotechnology Initiative (NNI), a federal initiative that includes many independent government agencies. The NNI began in 2000 and it has received an average of more than $1 billion dollars each year. From this, the NNI provides research funding to other agencies such as the National Science Foundation and Department of Defense. The goals of the NNI include the responsible development of a nanotechnology R&D program that can deliver new consumer and commercial products. It also promotes nanotechnology education that will train a workforce and build an infrastructure for future growth of nanotechnology.

The governments of Japan and the European Union currently support nanotechnology research and development at approximately $1 billion annually. China's funding is less but increases each year. These countries have goals similar to those of the NNI but may strike a different balance between basic research, technology transfer, workforce development, and safety.

Policy makers in many countries engaged with their citizens when nanotechnology was first established as a topic of national interest before many nano-enabled technologies reached maturity. This approach benefits researchers studying the underlying science, industries pursuing the use for practical applications, and government officials overseeing the R&D process. They can learn what issues concern the public the most then address them early in the technology development process. Educators can incorporate the new science and technology in courses so that students can immediately contribute to the new technology workforce. Consumers can indicate their needs for future products, which helps technology developers and manufacturers create the products that people want. Most importantly, early engagement builds trust with the public. A backlash due to the public's limited acceptance of risk becomes less likely, although not entirely avoidable. "The public" is neither a single entity nor does its members maintain a constant opinion as technologies evolve. An initial evaluation of nanotechnology risk is based on vague predictions about future applications and events. As products and services based on nanotechnology emerge, people may revise their own cost-benefit analysis. Decision making that accounts for a society's diverse and changing opinions is further complicated because effects may have a global scope. Weighing these competing interests creates ethical dilemmas that students can explore.

Another benefit is that a society can be aware of potentially radical change. This can occur if a single technology transforms not just a product or industry sector, but the entire culture. Examples of this include the automobile and cellular phones. While it may have been impossible to anticipate all the ramifications that these advancements, people would have benefited by having more time to plan for, understand, and adjust to the sudden changes in that took place in transportation and communications.

Early engagement of government, industry, and researchers with the public is referred to as upstream societal engagement and it is unfortunately an uncommon occurrence. In the past, governments would follow a downstream engagement approach. This involved policy makers setting national-level research and development priorities that allowed the development of certain new technologies without public consultation. Once these products reached their market, the public would become interested and begin a discussion with their leaders. At that point, it may be too late – important decisions about the technology and its use have already been made so it is difficult to fundamentally change how the technology is used in the future.

As part of their upstream engagement, national governments have established nanoELSI (nanotechnology-related ethical, legal, and societal issues) programs that investigate both beneficial and negative aspects. NanoELSI scholarship draws from diverse fields such as economics, philosophy, history, law, political science, and many others. Researchers have identified potential concerns in the areas of the environment, equity, laws and regulations, privacy, medical ethics, military applications, and human enhancement, as well as the rapid pace of advancement in these fields (hypertechnology).[4,5] These topics are the basis for many activities that can help students learn and develop their own views about nanotechnology and ethics.

8.3 Teaching Ethical Issues

Ethical considerations raised by nanotechnology R&D are like those of any science, or other human endeavours for that matter. Relationships between security and privacy, economic gain and environmental protection, or individuals and collective society, are at the heart of many nanotechnology debates. Rather than explore these in the abstract, this section presents ways for students to engage in ethical discussions surrounding nanotechnology. Some activities were originally designed for informal education involving the public or for college students in a nanotechnology class. Those activities presented here appear to be suitable for a middle or high school curriculum, perhaps with some modifications. Activities that highlight safety and environmental concerns appear in Chapter 5. As with the lab activities, readers are encouraged to contact the original author of the activity if it is not freely available.

8.3.1 Class Discussions

An example of upstream engagement is a consensus conference, in which a diverse group of lay people meet to learn about a topic, discuss the benefits and risks among themselves, and document their views. In the process, the participants learn about the technical and societal aspects of a topic such as nanotechnology. Their findings can inform experts and policy makers about previously overlooked issues. In an activity originally designed for middle school students but also adapted for an adult audience, a teacher can simulate a consensus conference among a class of students.[6] Each pair of students receives a packet of information about one nanotechnology research area. The two students read and discuss the information then agree on three benefits and three risks. When the whole class reassembles, each pair of students presented their arguments for the benefits and risks of their research area. Through several rounds of discussion and voting, the class determines the amount of funding that each research area should receive from a total budget of $1.5 billion. Adults completing this activity showed increases in their knowledge about nanotechnology. They also felt that they were well-informed and had increased confidence in their understanding of the issues. They indicated that they would be open to learning more about and discussing nanotechnology. This activity promoted learning about nanotechnology and its applications. Participants used that knowledge to form opinions about the risks and benefits as well as the relative importance of each research field. They collaborated and debated with others in a realistic budgeting exercise to distribute $1.5 billion for research funding, an amount equal to resent NNI budget requests. This activity would be appropriate for a science or civics curriculum.

Bionanotechnology – the application of nanotechnology to living systems – has great potential for alleviating human suffering. Still, bioethicists are concerned about these developments for many reasons, including the loss of patient privacy, unequal access to treatments and diagnostics, and environmental pollution. Doctors can now collect more information about patients thanks to the ability to diagnose illnesses in more detail. This data must be secure to protect privacy and there is debate about how governments, medical insurance companies, and the patients themselves would respond if they knew about likelihoods of disease, genetic disorders, *etc.* Often, new technologies are expensive, so they are available to only rich individuals and countries. In fact, a profit motive can determine which diseases are targeted for treatment in the first place. Companies favour first world illnesses that can allow them to recoup their research investment, while problems like tuberculosis are ignored because it mainly affects poorer countries. Nanomaterials such as silver nanoparticles (see Section 8.2.4) can treat infections by killing microbes but the unintentional release of those same particles can have a negative impact on the environment.

Middle and high school students can learn about these ethical issues in several activities. The first is a class discussion about bionanotechnology in general.[7] Resources include links to histories of inequitable and unethical medical research, such as the Tuskegee Syphilis Study. After the teacher guides students through a presentation about nanotechnology and ethics, they discuss with each other some of the issues raised by the presentation and online resources. Groups give a presentation to the class about a selected ethics topic and write a report. A second activity focuses on low-cost paper diagnostics used to diagnose AIDS.[8] A series of videos and presentations by the teacher give students information about this low-cost tool for detecting diseases, the engineering design process for these devices, and how paper diagnostics can improve health care in developing countries.

Students discuss some of the ethical issues then prepare a public service announcement (PSA) that informs their peers about AIDS and testing.

8.3.2 Games

The NanoVenture board game invites players to fund nanotechnology R&D on a global scale.[9] As leaders of different countries, players develop a strategy for investing in nanotechnology for industry, defense, and basic research. In addition to budgeting, players must keep their citizens satisfied. Decisions in the game have consequences. For instance, by purchasing LED lights for traffic signals, the country saves money on energy costs. However, the people of the country are unhappy because the use of LEDs causes job losses in the incandescent light bulb industry. Players can invest in new research that generates intellectual property, which other players must pay to use. Players' knowledge affects the game play, with rewards or lesser penalties given if players can answer questions about nanotechnology and its applications. Players are disqualified if they run out of money or their popularity within their country drops to 0%. The game structure is similar to Monopoly® with the winner being the person (country) with the most money or if all other players drop out. The short- and long-term ramifications for decisions and their impact on the leader's popularity adds another dimension of realism to this activity, as does the competition between countries.

The card game Decide – Nanotechnology leads students through a structured discussion of nanotechnology's potential benefits and problems.[10] The goals of the game are for participants to clarify their own individual views about nanotechnology then reach a consensus with other players. Instructions encourage respectful debate, sharing ideas, and finding common ground. As a group, players decide the topics they wish to discuss. The game provides four general policy positions, ranging from no research and implementation of nanotechnology until there is a sustained consensus among the public to enthusiastic embrace of nanotechnology with minimal regulations so that the public enjoys the benefits of nanotechnology as soon as possible.

8.3.3 Case Studies

Case studies can draw students into a rich narrative, presenting them with a controversy but no obviously correct resolution. Students must work that out for themselves, perhaps by playing the role of a character in the story. This can be liberating for students because they feel encouraged to think creatively and incorporate their own ideas into the story. Case studies based on technical data provided to students can be especially educational. Students are accustomed to performing expository laboratory experiments that are designed to produce a clear answer. Real data are often messy and incomplete, making it impossible to draw firm conclusions and exclude all other explanations. It can be frustrating for those who prefer a clear-cut, right answer. Students studying science and engineering may not be used to this level of ambiguity in class assignments.

Case studies are just one method for incorporating socioscientific issues (SSI) to their curriculum. Other approaches include debates, which have been shown to improve students' scientific competencies and attitudes towards SSI such as sustainability.[11] Regardless of the method for introducing SSI, there is a well-established need for teachers to connect scientific facts with students' real-world experiences.[12] Readers should consult the research literature to understand how students respond to SSI activities[13] and to find current examples of SSI activities to use in their classes.

Historical and fictional case studies each have advantages and disadvantages. Case studies based on actual events reinforce the sense of realism that draws students into the story. They can be easy to find because, unfortunately, negative outcomes from poorly decided ethical choices appear in the news frequently. As a disadvantage, students can find information about actual events. Although history will not necessarily describe the best possible outcome, it can bias students' decisions if they know what actually happened. Fictional scenarios give the teacher greater leeway in constructing a story but creating a compelling narrative requires more work. Topics can be customized for the particular course material, degree of difficulty, student interests, or geographical location. Regardless of which approach a teacher takes, it is important that students play the role in a way that is generally consistent with the desires and motivations of the character, rather than their own opinions. It is not necessary that the student agree with their character's viewpoint. Case studies will need periodic updating as the scenarios become outdated or future nanotechnology applications change.

An example of a nanotechnology case study based on real events is the publication of falsified results of molecular electronics research at Bell Labs.[14] Obviously, the scientist who committed this deception, Henrik Shon, was unethical. More interesting are the decisions made by Shon's colleagues and supervisors that ultimately allowed the falsified results to be published. In a case study like this, students can learn about events and circumstances that occurred behind the scenes. The temptation to publish questionable but ground-breaking research is real and not unique to this instance. Reduced funding for research, increased competition among research groups, the need for companies to be immediately profitable, and the desire for recognition as a great scientist can lead otherwise decent people to engage in unethical, or at least questionable behaviour. By addressing challenges like this, the students can learn to be more ethical scientists.

Another example of a case study activity for high school students is modelled after a U.S. television show "Nanotechnology: The Power of Small".[15] The original three-part program consisted of a moderator and a panel of experts in various fields of science journalism, research, and policy. They discussed realistic but fictional situations that involve near-term nanotechnology applications. Each case study revolves around a group of characters role played by members of the panel who must decide if or how they can use nanotechnology to achieve their desired goals. An example of this format can be found in other seminar videos.[16] Program segments deal with nanomedicine, nanoscale electronics, and environmental impacts.

The author performed a version of this activity with first-year college students and high school students interested in nanotechnology.[17] Power of Small case studies were modified to make them more relevant to younger students. Whenever possible, details like real places, events, and people are used. RFID chips allowed local businesses to track students' shopping habits in their stores in return for discounts and product recommendations. A student-led company wanted to manufacture nanothin films on campus but would not disclose their materials to protect intellectual property. Campus police wanted to use nanoelectronics for monitoring students to prevent crime. Additional scenario ideas ranging from the environmental impact of antimicrobial socks to human enhancement can be found online.[18] Each participant is given a backstory for the role of a person involved in each scenario. During the panel discussions, characters are graded on their ability to support their views with data and evidence. After the activity, students write position papers about their scenario based on their own opinion, not the opinion of their character. The teacher acts as a moderator who can guide the discussion and keep all panellists involved in the discussion.

Students enjoyed this activity because they could augment the basic character background with their own details. Some students dressed in costume and demonstrated strong acting skills. Each year, the discussions would follow a different course based to the ability of students to persuade their peers. In most cases, the panel could reach a compromise but that was not required. Interestingly, the high school and college students brought up the same ethical concerns that the experts in the Power of Small show did. While these case studies can, on the surface, appear to be relevant to just nanotechnology, the underlying ethical issues are universal: safety, privacy, environmental protection, personal autonomy, and economics.

8.3.4 Media-based Activities

Entertainment media can provide middle and high school students with many examples of near- and distant future uses of nanotechnology. These presentations more often show nanotechnology as dangerous or at least highlighting ethical questions about its applications. Teachers should provide background information about nanotechnology and which parts of the media accurately describe current developments of nanotechnology. It is important that teachers make clear which aspects of the stories are likely impossible to achieve so that the activity does not create misconceptions. Using television and movies introduces the problem of keeping the material current since students might not recognize a movie that is older than they are. Some media may not be appropriate for younger students.

Several activities describe how teachers can use movies and other media for introducing ethical issues of nanotechnology. The first describes a general approach for teaching about nanotechnology, sharing the media with students, and leading a discussion.[19] Resources are provided for each of these steps, although the website and media lists are out of date. Teachers should search online for more a more current catalogue.

Science fiction literature provides examples of how people can resolve moral and ethical conflicts created by nanotechnology. Two examples of stories about nanotechnology affecting society are Michael Flynn's *The Nanotech Chronicles*[20] and Neal Stephenson's *The Diamond Age*.[21] Teachers can assign these readings to students for reflection about how nanotechnology will affect society. The activity can be a writing assignment[22] or class discussion.[23] (ref. 22 lists many other nanotechnology-themed stories. Ref. 23 contains specific prompts for discussion of ethics drawn from the story.) Science fiction stories give students one example of how nanotechnology can change society. This can prompt them to imagine alternative versions, better and worse. They can then consider the choices that people must make to achieve a preferred outcome.

8.3.5 Responsible Research and Innovation and Nanotechnology: An Activity that Incorporates Multiple Approaches

Responsible Research and Innovation (RRI) is a policy framework initially developed by the European Union that connects science to six larger societal issues: engagement among stakeholders, gender equality in science, science education, open access to research results, ethics, and governance. It has been useful for helping scientists and policy makers build trust with the public about current and future nanotechnology research and applications. Teachers can use the aspects of RRI to introduce students to nanotechnology.

An activity was developed for sixth grade students to learn about size and scale and size dependent properties of nanomaterials.[24] This was an inquiry-based, multi-session unit. The first four lessons introduced nanotechnology concepts. Students viewed a video about current and near-future nanotechnology applications, such as self-cleaning surfaces, then reflected on how those materials might affect their daily lives. They visited a science museum and discussed how well exhibits conveyed ideas about nanotechnology. Students performed a series of measurements that impressed upon them the smallness of the nanoscale and illustrated the relative sizes of objects. The next lesson focused on size dependent properties as they relate to hydrophobic and hydrophilic surfaces. This provides students with an explanation for the performance of self-cleaning surfaces that were introduced in the first lesson. Students worked with models to learn about the change in surface to volume ratio as particle size decreases.

The latter three lessons required students to apply their new knowledge of nanotechnology. Students read and discussed news articles about nanotechnology and considered how the articles reflected aspects of RRI. The class interviewed a nanotechnology expert about ethical issues and the role of scientists in the development of new applications. For the final lesson, groups of students prepared exhibits about nanotechnology.

Researchers found that this series of lessons helped younger students better understand the size and scale of nanosized objects and the size dependent properties at the nanoscale. Students better understood the issues related to RRI dimensions of engagement among stakeholders, science education, and ethics.

8.4 Environmental, Health, and Safety Concerns of Nanomaterials

Nanomaterials appear in a wide variety of consumer goods that come into close contact with people, including foods, cosmetics, and textiles. They are used in the production of other materials as well, such as nanocatalyst particles for chemical manufacturing. This raises questions about their impact on the environment, health, and safety (EHS). Exposure to nanomaterials during their production and their use in manufacturing can harm the health of workers. Contact with nanomaterials in products can impact the health of consumers. Disposal of products containing nanomaterials can affect workers. Nanomaterials can enter an ecosystem during all three of these phases of a product's life cycle. Laboratory experiments described in Chapter 5 illustrate why people are concerned about the impact of nanomaterials on health and the environment.

In 2012, worldwide production of nanomaterials was estimated to be about 10 000 metric tonnes per year with nanosized silicon dioxide annual production accounting for over half of that total.[25] Nanomaterial production is almost certainly greater ten years later. The most common nanomaterials are, in decreasing order: SiO_2, TiO_2, ZnO, carbon nanotubes, silver, AlO_x, FeO_x, CeO_x, fullerenes, and other quantum dots. This only addresses engineered nanomaterials; naturally occurring nanomaterials are even more plentiful. Sources include combustion that form carbonaceous nanoparticles in soot and weathering of rocks.

Degradation of discarded plastics is another source of nanoparticles and is easily demonstrated to students in a nanomaterials safety workshop.[26] After learning fundamental concepts of nanotechnology (*e.g.*, nanoscale sizes, surface to volume ratio), the instructor disintegrates a piece of polystyrene in water using a consumer-grade blender. The water and polystyrene mixture is passed through a syringe with a 0.8 micron (800 nm) filter. The path of a laser pointer beam passing through the filtered liquid is visible, indicating that sub-micron sized polystyrene particles are present. See the description of the Tyndall effect in Chapter 1, Nanotechnology and Chemistry: The Unseen Scale with Magnificent Impact for an explanation of this phenomenon. Students perform other activities to observe that organisms interact with the nanoparticles in the environment.

Students can learn about the presence of nanomaterials in common products by researching their production and reading news articles.[27] From that starting point, students can learn about any EHS concerns for a particular type of nanomaterial. However, its EHS impact will depend on many variables besides the chemical formula and dimensions, as explained below. This information is often not readily available to consumers. To further complicate consumers' ability to understand the risks of a product, some products claim to contain nanoparticles when in fact they do not while other products conceal the presence of nanoparticles. Requirements for companies to disclose nanomaterial content vary by country.

Besides their presence in consumer products, the method of manufacturing the nanomaterial influences its EHS effects. The most common method is top-down production. In this process, a bulk material is broken down into smaller units, followed by separation of the desired particle sizes. This generates more waste than bottom-up manufacturing, in which atomic scale units of a material are assembled until the desired particle sizes are achieved.

The field of nanotoxicology is concerned with the negative effects of nanomaterials on organisms and their environment. Traditionally, toxicity depends on the dose of a chemical, its chemical properties, and the method of exposure (*e.g.*, inhalation). Nanomaterials introduce additional variables which make researchers' ability to understand the EHS issues of current nanomaterials and predict problems arising from future nanomaterials more difficult.

Obviously, size plays a key role in determining the properties of nanomaterials. As the size of particles decrease, they are more reactive and their usual bulk properties change as well. Nanoparticles are small enough to attach themselves to the outer membrane of a biological cell and potentially enter the cell through endocytosis, diffusion, or other mechanisms. The greater reactivity of the nanoparticle's surface atoms and ions can lead to undesirable reactions within the cell, causing mutation or cell death. On the other hand, the higher surface reactivity of smaller nanoparticles tends to promote aggregation, forming more stable, larger particles. The degree of aggregation depends greatly on the chemical environment around the nanoparticles.

Nanomaterial shape plays a large role in its toxicity as well. Carbonaceous nanomaterials provide a ready example of this. Round fullerenes, flat sheets of graphene, single cylinders of graphene forming single walled carbon nanotubes, and multiple layers of rolled graphene as multi-walled carbon nanotubes are all composed entirely of carbon but exhibit different chemical and biological behaviour. Spherical nanoparticles tend to be absorbed more easily than other shapes into a cell. However, longer carbon nanotubes are more likely than shorter nanotubes to become lodged in lung tissue, leading to inflammation.

As demonstrated in the experiments in Chapter 5, nanoparticle properties change with their surface coating. For example, surface-bound citrate ions stabilize gold nanoparticles. Adsorbed molecules or ions can affect the nanoparticle solubility, which can greatly affect its cellular uptake. A coating applied during the manufacturing process may be stable for as long as the product is to be used but once discarded, the different chemical environment may cause that coating to desorb and be replaced by another, changing the reactivity of the nanomaterial.

8.5 Conclusions

It is clear that nanomaterials possess great promise for improving people's quality of life, but those benefits are not always accessible to all, and the negative EHS impacts are just as obvious. As part of a well-rounded education and preparation for functioning in a global society, students must begin to consider how to balance the many ethical issues related to nanotechnology. Fortunately, educational activities exist to guide them in this process.

References

1. I. A. Joubert, M. Geppert, S. Ess, R. Nestelbacher, G. Gadermaier, A. Duschl, A. C. Bathke and M. Himly, *NanoImpact*, 2020, **17**, 1.
2. NNIN Education Program, https://www.nnin.org/education-training/k-12-teachers/nanotechnology-curriculum-materials/exploring-nanotechnology, (accessed May 2021).
3. NANOYOU Nano Dialogue, https://nanoyou.eu/en/decide.html, (accessed May 2021).
4. B. Baird and T. Vogt, *Nanotech. Law Bus. J.*, 2004, **1**, 101.
5. F. Alloff, P. Lin and D. Moore, *What is Nanotechnology and Why Does It Matter?* Wiley-Blackwell, Sussex, 2010.
6. A. R. Jones, A. A. Anderson, S. K. Yeo, A. E. Greenbert, D. Brossard and J. W. Moore, *J. Chem. Educ.*, 2014, **91**, 179.
7. Help or Hype: The Ethics of Bio-nanotechnology, https://www.nnin.org/education-training/k-12-teachers/nanotechnology-curriculum-materials/help-or-hype-ethics-bio, (accessed May 2021).
8. Paper Diagnostics, https://www.nnin.org/education-training/k-12-teachers/nanotechnology-curriculum-materials/paper-diagnostics, (accessed May 2021).
9. NanoVenture: The Nanotechnology Board Game, https://chem.beloit.edu/edetc/supplies/nanoventure, (accessed May 2021).
10. Play DECIDE - Nanotechnology https://nanoyou.eu/en/component/content/article/11-cards/88-play-decide-nanotechnology-education-resources1e2d.html, (accessed May 2021).
11. C. Y. Tsai, *Comput. Educ.*, 2018, **116**, 14.
12. A. Hofstein, I. Eilks and R. Bybee, *Int. J. Sci. Math. Educ.*, 2011, **9**, 1459.

13. S. C. Fang, Y. S. Hsu and S. S. Lin, *Int. J. Sci. Math. Educ.*, 2019, **17**, 427.
14. M. E. Gorman, N. Swami, C. J. McGrath, J. Groves, K. Squibbs and P. H. Werhane, *J. Nano Educ.*, 2013, **4**, 25.
15. Nanotechnology: The Power of Small, https://www.wilsoncenter.org/article/nanotechnology-the-power-small, (accessed May 2021).
16. Minds on Edge: Facing Mental Illness A Fred Friendly Seminar, https://youtu.be/G6PpSmTlyRY, (accessed May 2021).
17. K. Winkelmann, *J. Nano Educ.*, 2013, **4**, 67.
18. NANOYOU: Role Play, https://nanoyou.eu/en/component/content/article/12-role-play/90-nanoyou-role-play-nanotechnology-education-resources1e2d.html, (accessed May 2021).
19. Using Media to Explore Social and Ethical Issues in Nanoscience and Nanotechnologies, https://www.nnin.org/education-training/k-12-teachers/nanotechnology-curriculum-materials/using-media-explore-social-and, (accessed May 2021).
20. M. Flynn, *The Nanotech Chronicles*, Baen Books, Riverside, NJ, 1991.
21. N. Stephenson, *The Diamond Age; or, A young Lady's Illustrated Primer*, Bantam, New York, 1995.
22. Reading and Analyzing Nanotechnology, https://www.nnin.org/education-training/k-12-teachers/nanotechnology-curriculum-materials/reading-and-analyzing, (accessed May 2021).
23. R. W. Berne and J. Schummer, *Bull. Sci. Technol. Soc.*, 2005, **25**, 459.
24. A. Mandrikas, E. Michailidi and D. Stavrou, *Res. Sci. Technol. Educ.*, 2020, **38**, 377.
25. F. Piccinno, F. Gottschalk, S. Seeger and B. Nowack, *J. Nanopart. Res.*, 2012, **14**, 1109.
26. T. Cedervall, M. T. Ekvall, K. Mattsson and M. Lundqvist, *J. Chem. Educ.*, 2019, **96**, 1967.
27. Exploring nanotechnology through Consumer Products, https://www.nnin.org/education-training/k-12-teachers/nanotechnology-curriculum-materials/exploring-nanotechnology, (accessed May 2021).

Subject Index

Illustrations (figures and tables) are comprehensively referred to in the text. Therefore, significant material in them has only been given a page reference in the absence of its concommitant mention in the text referring to those illustrations.